Hardware/Software Co-Design and Optimization for Cyberphysical Integration in Digital Microfluidic Biochips

Yan Luo • Krishnendu Chakrabarty • Tsung-Yi Ho

Hardware/Software Co-Design and Optimization for Cyberphysical Integration in Digital Microfluidic Biochips

 Springer

Yan Luo
Department of Electrical
    and Computer Engineering
Duke University
Durham, NC, USA

Krishnendu Chakrabarty
ECE, Duke University
Durham, NC, USA

Tsung-Yi Ho
CS & Info Engineering
National Cheng Kung University
Tainan City, Taiwan

ISBN 978-3-319-35993-9   ISBN 978-3-319-09006-1 (eBook)
DOI 10.1007/978-3-319-09006-1
Springer Cham Heidelberg New York Dordrecht London

*To my beloved grandfather and grandmother,
Junyi Luo and Fangzhen Wang — Yan Luo*

Preface

Microfluidic biochips have now come of age, with applications to biomolecular recognition for high-throughput DNA sequencing, immunoassays, and point-of-care clinical diagnostics. In particular, digital microfluidic biochips, which use electrowetting-on-dielectric to manipulate discrete droplets (or "packets of biochemical payload") of picoliter volumes under clock control, are especially promising. The potential applications of biochips include real-time analysis for biochemical reagents, clinical diagnostics, flash chemistry, and on-chip DNA sequencing. The ease of reconfigurability and software-based control in digital microfluidics has motivated research on various aspects of automated chip design and optimization.

This book is focused on facilitating advances in on-chip bioassays, enhancing the automated use of digital microfluidic biochips, and developing an "intelligent" microfluidic system that has the capability of making on-line re-synthesis while a bioassay is being executed. This book includes the concept of a "cyberphysical microfluidic biochip" based on the digital microfluidics hardware platform and on-chip sensing technique. In such a biochip, the control software, on-chip sensing, and the microfluidic operations are tightly coupled. The status of the droplets is dynamically monitored by on-chip sensors. If an error is detected, the control software performs dynamic re-synthesis procedure and error recovery.

In order to minimize the size and cost of the system, a hardware-assisted error-recovery method, which relies on an error dictionary for rapid error recovery, is also presented. The error-recovery procedure is controlled by a finite-state machine implemented on a field-programmable gate array (FPGA) instead of as software running on a separate computer. Each state of the FSM represents a possible error that may occur on the biochip; for each of these errors, the corresponding sequence of error-recovery signals is stored inside the memory of the FPGA before the bioassay is conducted. When an error occurs, the FSM transitions from one state to another, and the corresponding control signals are updated. Therefore, by using inexpensive FPGA, a portable cyberphysical system can be implemented.

In addition to errors in fluid-handling operations, bioassay outcomes can also be erroneous due to the uncertainty in the completion time for fluidic operations. Due to the inherent randomness of biochemical reactions, the time required to complete each step of the bioassay is a random variable. To address this issue, a new "operation-interdependence-aware" synthesis algorithm is described in this book. The start and stop times of each operation are dynamically determined based on feedback from the on-chip sensors. Unlike previous synthesis algorithms that execute bioassays based on pre-determined start and end times of each operation, the proposed method facilitates "self-adaptive" bioassays on cyberphysical microfluidic biochips.

Another design problem addressed in this book is the development of a layout-design algorithm that can minimize the interference between devices on a biochip. A probabilistic model for the polymerase chain reaction (PCR) is described; based on the model, the control software can make on-line decisions regarding the number of thermal cycles that must be performed during PCR. Therefore, PCR can be controlled more precisely using cyberphysical integration.

To reduce the fabrication cost of biochips, yet maintain application flexibility, the concept of a "general-purpose pin-limited biochip" is presented. Using a graph model for pin-assignment, the book develops the theoretical basis and a heuristic algorithm to generate optimized pin-assignment configurations. The associated scheduling algorithm for on-chip biochemistry synthesis has also been developed. Based on the theoretical framework, a complete design flow for pin-limited cyberphysical microfluidic biochips is presented.

In summary, this book presents an algorithmic infrastructure as well as a set of optimization tools for cyberphysical system design and technology demonstrations. The results presented here are expected to enable the hardware/software co-design of a new class of digital microfluidic biochips with tight coupling between microfluidics, sensors, and control software.

The authors acknowledge Professor Bhargab B. Bhattacharya from Indian Statistical Institute, Kolkata for valuable discussions and collaboration. The authors also acknowledge Professor Richard B. Fair, Andrew C. Madison, Dr. Bang-Ning Hsu, Liji Chen, and Dr. Yan-You Lin from Duke University for valuable discussions and collaboration.

Yan Luo and Krishnendu Chakrabarty acknowledge the financial support received from US National Science Foundation.

Durham, NC, USA
Durham, NC, USA
Tainan City, Taiwan

Yan Luo
Krishnendu Chakrabarty
Tsung-Yi Ho

Contents

Chapter 1
Introduction

Microfluidic biochips have emerged as powerful and reliable toolkits for biotechnology applications, such as chemical synthesis, the diagnosis of diseases, and the development of new drugs [1–3]. Nanoliter and picoliter volumes of biological samples can be manipulated on microfluidic devices under software control. Compared to conventional devices and analyzers, microfluidic devices offer many unique advantages. The low volume of samples and reagents manipulated on microfluidic devices can significantly reduce the costs associated with conducting experiments; the precise control of reactions on microfluidic devices can enhance the accuracy of the experiments; the time required for the chemical reactions to occur at the nanoliter scale can be greatly reduced due to the high surface-to-volume ratios. Based on the methods used to manipulate the liquid on the chip, microfluidic biochips are categorized as "flow-based chips" or "digital (droplet-based) chips" [4, 5].

In continuous-flow chips, liquid flow on the biochip is achieved through micro-fabricated channels, pumps, and valves [6, 7]. To overcome fluidic resistance, liquid in the channels is driven by external pressure sources. Such sources include external mechanical pumps and integrated, mechanical micropumps. The pressure required for liquid flow can also be provided by combinations of capillary forces and electrokinetic mechanisms [6]. In recent years, the development of flow-based microfluidic devices has been accelerated by innovations in fabrication techniques, including the application of polydimethylsiloxane (PDMS) and the dense integration of active microvalves. From the early days, when flow-based biochips had simple topologies and only a few channels, commercial flow-based microfluidic devices today have large-scale networks of channels, and they can be used in various real-world applications. For example, a product from Fluidigm [8], a biotechnology company that focuses on flow-based microfluidic biochips, can perform a series of gene analyses, including enrichment for target DNA sequences, sample barcoding for multiplexed sequencing, and preparation of the sequencing library.

© Springer International Publishing Switzerland 2015
Y. Luo et al., *Hardware/Software Co-Design and Optimization for Cyberphysical Integration in Digital Microfluidic Biochips*, DOI 10.1007/978-3-319-09006-1_1

However, flow-based biochips have several inherent limitations and drawbacks. First, flow-based chips are difficult to integrate and scale down, as the scale of each component in the device may have a decisive impact on the performance of the entire system. The second problem is that these chips cannot be reconfigured after they are fabricated because the channels are etched in the substrate [4]. Thus, the paths for liquid flow and all of the operations implemented by the chip are predetermined, and no software-based differentiation is possible. The third problem is that these chips lack fault tolerance, hence the entire chip ceases to be functional if any channel or other on-chip element is defective [4].

Digital (droplet-based) microfluidic biochips have been proposed as an alternative to flow-based microfluidics [4, 5]. On the droplet-based microfluidic platform, liquid droplets and all of the fluid-handling operations are manipulated using an array of discrete electrodes [1, 3]. Since each droplet is analogous to "a bit of information" and operates under clock control, this device is referred to as a "digital microfluidic biochip" [3].

In digital microfluidic biochips, all molecular processes and biochemical reactions are conducted using discrete droplets (or "packets of biochemical payload") that have nanoliter/picoliter volumes, which allows a very significant reduction in reagent/sample volume and reaction time. As microfluidic droplet platforms have the capability of conducting fast and efficient mixing inside droplets, high throughput can be achieved for experiments. The droplets on digital microfluidic biochips have no contact with any of the solid walls of the flow channels because they are surrounded by silicone oil and manipulated on the surface of the electrode array. Thus, as a benefit of this structure, the risk of cross contamination in experiments and the adsorption of reagents/samples by the walls of the channels are minimized. Therefore, the quality of the product droplet and the reliability of the biochip are improved.

Since discrete droplets are manipulated independently on the biochip, the droplet-based microfluidic biochip is especially suitable for researchers to closely monitor various reactions over time. For example, researchers have developed a droplet-based process for the identification and enumeration of foodborne pathogens [9]. In this process, each water-in-oil droplet acts as a microreactor for the encapsulation of the cell. By observing the metabolic activity of a single bacteria cell that is confined by a picoliter-scale droplet, the detection and enumeration of bacteria can be performed quickly and precisely. The oil that surrounds each droplet acts as a carrier of droplets and a barrier against cross contamination. Compared with the conventional methods used to detect pathogens, in which researchers have to incubate the specimens for several hours or even days before the pathogens can be detected by conventional instruments, methods based on digital microfluidic biochips can significantly reduce detection time and improve the reliability of the measurement results.

Given all these advantages, digital microfluidics is now seeing increasing acceptance in biotechnology applications that require high-precision control of fluid flow during experiments [1–3]. The complexity of such systems and the integration level of digital microfluidic biochips has increased markedly in the last few years. For example, a commercially-available droplet-based microfluidic system embeds more than 300,000 electrodes with integrated optical detectors [10]. The system has been developed and optimized for protein expression, single-cell gene expression, single nucleotide polymorphism (SNP) genotyping, and the quantification of samples of DNA strands.

As the applications of digital microfluidic biochips grow, greater demands are likely to be placed on the quality of product droplets and the accuracy of fluid-handling operations. At the same time, the risk of errors during the execution of a bioassay is likely to increase due to the greater number of on-chip operations. For example, if a larger droplet that is to be split is not placed at the center of the splitter, unbalanced splitting may occur, and the resulting smaller droplets will have erroneous volumes.

Unlike the detection of defects involving the electrodes and the breakdown of the insulation layer, operational errors are difficult to predict or detect before the bioassay is conducted. For a biochip that does not have any hardware defects, operational errors are the main cause for the failure of bioassays. In order to improve the yield of bioassays and the automation of digital microfluidic biochips, it is necessary to develop an "intelligent" microfluidic system that has the capability of making on-line re-synthesis while a bioassay is being conducted.

This book is focused on a new design and development strategy for cyberphysical microfluidics-based biochips that are suitable for large-scale bioassay applications. Based on the existing hardware platform of biochips and emerging on-chip sensing techniques, the concept of a cyberphysical microfluidic biochip is proposed in this book. In such a biochip, the control software and the hardware platform are tightly coupled. The status of the droplets is monitored in real-time by on-chip sensors. If an error is detected, the control software performs dynamic resynthesis to determine how to recover from the error. The dynamic re-synthesis procedure reschedules operations and reallocates on-chip resources to operations in order to minimize the completion time of the bioassay, while guaranteeing that there is no conflict in resource sharing. By applying the control signal sequences derived by the re-synthesis procedure, the biochip discards the faulty droplets immediately to stop the propagation of the error. The operation that generated the faulty droplet(s) will be repeated. In this way, the biochip recovers from the effects of the operational error.

In order to minimize the size and cost of the system, a hardware-assisted error recovery method is also discussed in this book. Instead of software running on a computer, a finite-state-machine (FSM) implemented on a field-programmable gate array (FPGA) is used to control the error-recovery procedure. Each state of the FSM represents one possible error that may occur on the biochip; for each of these errors, the corresponding sequence of error-recovery signals is stored in the memory of the FPGA before the bioassay is conducted. When an error occurs, the FSM transitions from one state to another, and the corresponding control signal is updated. Therefore, by using simple and inexpensive hardware (the FPGA), a portable, cyberphysical system can be implemented. This design is particularly convenient for handheld or portable devices that may emerge with future advances in technology.

In addition to errors in fluid-handling operations, another possible cause for the failure of a bioassay is the uncertainty in the completion time of fluid-handling operations. Due to the inherent randomness of biochemical reactions, the time required to complete each step of the bioassay can be viewed as a random variable. To address this issue, a new "operation-interdependence-aware" synthesis algorithm is proposed in this research work. The start and stop time of each operation are determined by feedback from the on-chip sensor. Bioassays can therefore be carried out in a self-adaptive fashion on cyberphysical microfluidic biochips. With such a synthesis approach, the reliability and flexibility of the biochips are improved, and the applications of biochips can be extended beyond well-calibrated bioassays to the exploration of new biochemical protocols.

To reduce the cost of fabricating biochips while maintaining flexibility with respect to potential applications, the concept of a general-purpose pin-limited biochip is introduced in this book. Using a graph model for pin-assignment, we have constructed the theoretical basis and a heuristic algorithm to generate optimized pin-assignment configurations. The associated scheduling algorithm for on-chip biochemistry synthesis is also presented. A complete design flow for pin-limited cyberphysical microfluidic biochips is presented on the basis of these results.

Another design problem addressed in this book is a layout-design algorithm that can minimize the interference between the devices on a biochip. A probabilistic model for the polymerase chain reaction (PCR) process is considered; based on the model, the control software can make on-line decisions regarding the number of thermal cycles that must be performed in the PCR procedure. Therefore, the duration of the PCR can be precisely controlled.

The rest of this chapter is organized as follows. Section 1.1 presents an overview of digital microfluidics. Section 1.2 introduces the design automation aspects of digital microfluidic biochips. Section 1.3 presents a chapter-by-chapter outline for this book.

1.1   Overview of Digital Microfluidics

In this section, we present an overview of digital microfluidics, including the theory of electrowetting-on-dielectric, the hardware platform, sensing techniques, and fault models for biochips.

1.1.1   Theory of Electrowetting-on-Dielectric

The electrowetting effect has been defined as "the change in solid-electrolyte contact angle due to an applied potential difference between the solid and the electrolyte" [11]. A detailed analysis of this phenomenon is presented in this section.

Fig. 1.1 Side view of the setup for observation of electrowetting on dielectric phenomenon

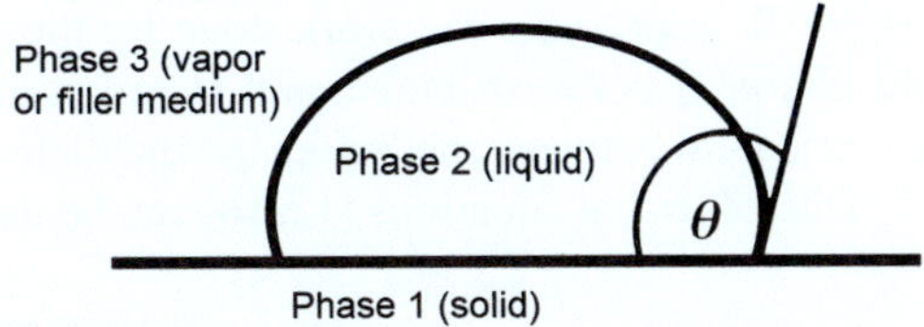

In the electrowetting-on-dielectric setup shown in Fig. 1.1, there are three phases that are in contact with each other [12]. Phase 1 is the solid dielectric layer; Phase 2 is the droplet. If the droplet is in direct contact with air, then Phase 3 is the vapor; if the droplet is surrounding by the filler medium, such as silicone oil, then Phase 3 is the surface of the filler medium. If no voltage is applied to the droplet, it will maintain its equilibrium shape. The contact angle between the phase of the droplet and the solid insulator is referred to as θ, which is defined as shown in Fig. 1.1.

Each interface has an interfacial energy, referred as γ_{ij}, where i and j are the indices that represent the corresponding phases. ***Interfacial Energy*** is defined as "the amount of energy, G, that must be used under reversible and isothermal conditions to increase the surface A of an interface at constant volume" [12–15]. The mathematical expression of this definition is:

$$\gamma_{ij} = \left. \frac{dG}{dA} \right|_{T=const, V=const} \tag{1.1}$$

According to ***the Principle of Minimum Total Potential Energy***, the total free energy of a droplet (G) should obtain its minimum value at the equilibrium state [12–15]. From Eq. (1.1), we obtain the basic correlation between the interfacial energy γ, the total free energy G, and the surface A, as:

$$\frac{dG}{dA} = \gamma_{sl} - \gamma_{sv} + \gamma_{lv} cos\theta = 0 \tag{1.2}$$

where the subscripts "l", "s", and "v" refer to liquid, solid, and vapor, respectively. Thus γ_{sl} represents the interfacial energy for the interface between the solid phase and the liquid phase. The symbols γ_{sv} and γ_{lv} are defined in a similar way. From the above equation, we can get the ***Young equation***, which shows that the value of the contact angle is a function of the interfacial energies [6, 12–15]:

$$cos\theta = \frac{\gamma_{sv} - \gamma_{sl}}{\gamma_{lv}} \tag{1.3}$$

When a voltage is applied to the droplet, the total free energy will contain additional terms [6, 12]:

$$dG = \gamma_{sl} dA - \gamma_{sv} dA + \gamma_{lv} dA cos\theta + dU - dW \tag{1.4}$$

where W represents the work done by the voltage source with voltage V when the charge q is redistributed, and U represents the energy required to establish an electric field between the liquid and the dielectric insulation.

The following equations [12, 16] can be used to calculate U:

$$U = \int_{Vol} \vec{E}\,\vec{D}\,dVol \tag{1.5}$$

$$\frac{U}{A} = \int_0^d \frac{1}{2}\vec{E}\,\vec{D}\,dz = \frac{1}{2}d\,\vec{E}\,\vec{D} \tag{1.6}$$

where $\vec{E}$ and $\vec{D}$ denote the electric field and the electric displacement, respectively.

$$\frac{dU}{dA} = \frac{1}{2}d\frac{V}{d}q \tag{1.7}$$

where V_{ol} represents the volume of the droplet, and d represents the thickness of the dielectric insulation layer on the surface of the electrode. In addition to the above equations, we also have:

$$\frac{dW}{dA} = Vq \tag{1.8}$$

and **Gauss's law**[16]:

$$q = \frac{\varepsilon_0\varepsilon_r V}{d} \tag{1.9}$$

Now, we can search for the minimum value of total free energy, G, by setting the derivative $\frac{dG}{dA}$ equal to zero, which results in the following equation [6, 12]:

$$\frac{dG}{dA} = \gamma_{sl} - \gamma_{sv} + \gamma_{lv}cos\theta - \frac{1}{2}\frac{\varepsilon_0\varepsilon_r}{d}V^2 = 0 \tag{1.10}$$

According to Eq. (1.10), we can express $cos\theta$ as the function of voltage applied on the electrode. Thus we get the **Young–Lippman Equation**:

$$cos\theta(V) = \frac{\gamma_{sv} - \gamma_{sl}}{\gamma_{lv}} + \frac{1}{2}\frac{\varepsilon_0\varepsilon_r}{\gamma_{lv}d}V^2 \tag{1.11}$$

Equation (1.11) shows that, when voltage is applied to the electrode, the solid-electrolyte contact angle can be adjusted by applying voltage to the droplet [17–19]. On the digital microfluidic platform, when control voltage is applied on the electrode, the surface of the insulator will change from hydrophobic to hydrophilic [17]. Thus, the droplet is repelled by the hydrophobic surface and attracted by the hydrophilic surface [17]. Due to the effect of this unbalanced surface tension, the droplet can be moved by applying a control voltage to the electrodes [17, 20].

1.1.2   Hardware Platform

A digital microfluidic biochip consists of a two-dimensional electrode array and on-chip reservoirs [20]. By utilizing the effect of electrowetting, nanoliter droplets containing biological/chemical samples and reagents can be manipulated on the chip without curved channels or external pressure sources [4, 17].

Figure 1.2 [17] shows the structure of a unit cell on a digital microfluidic biochip. The upper plate is a large electrode which covers all cells on the array and serves as the ground electrode for all unit cells [17, 20]. When the biochip is used, the upper plate is applied a common voltage, thus all the unit cells in the array have the same voltage on their upper electrodes [17, 20]. The lower plate of the unit cell consists of an array of discrete control electrodes. During chip operation, the unit cells in the array may have different voltages on their lower electrodes. The movements of droplets are determined by signals applied on the discrete electrodes. In the literature on microfluidic biochips, the term "control voltages applied to electrodes" usually refers to the voltages applied to the lower electrodes of the unit cells on the chip.

As shown in Fig. 1.2 [17], droplets manipulated by the digital microfluidic biochip are confined between the upper and lower electrodes [17]. To move a droplet, a high voltage should be applied to a unit cell adjacent to the droplet, and at the same time, a low voltage must be applied to the cell under the droplet [17]. The voltages applied to the electrodes can influence the surface characteristics of the hydrophobic coating on the lower electrodes. In this way, different voltages applied to the electrodes result in multiple levels of interfacial tension on the surface of the biochip [17]. Due to this effect, the droplet is moved from the low-voltage electrode to the high-voltage electrode[17]. This is the basic operating principle of digital microfluidic biochips. All microfluidic functions, such as droplet merging, splitting, mixing, and dispensing can be reduced to a set of basic operations [17]. Concurrent manipulation of multiple discrete droplets can be coordinated by control software and by voltages applied to the electrodes, without the use of mechanical devices, such as tubes, pumps, and valves [4].

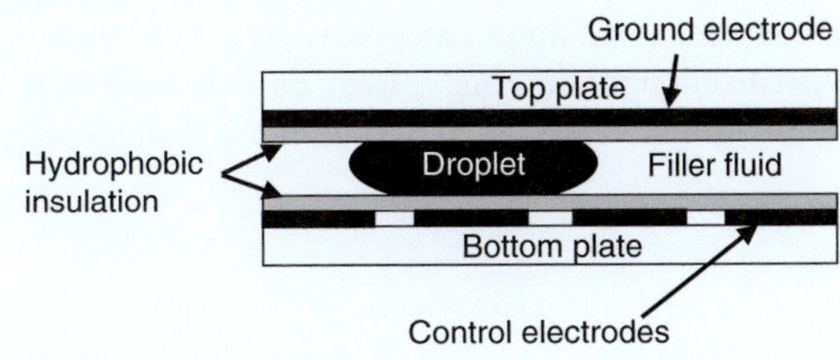

Fig. 1.2 Schematic cross-section of a unit cell on the microfluidic biochip [17]

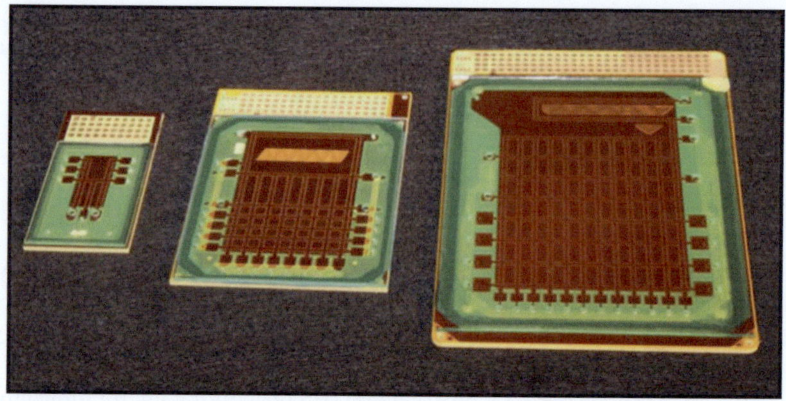

Fig. 1.3 Several generations of fabricated and packaged digital microfluidic biochips [21]

Figure 1.3 [21] shows several generations of fabricated and packaged digital microfluidic biochips. The electrodes in these devices are fabricated on printed circuit boards (PCBs). Similar biochips have been fabricated on glass and silicon, and demonstrated for a wide variety of biomedical assays, including on-chip chemistry for DNA sequencing [1], multiplexed real-time polymerase chain reaction [22], protein crystallization for drug discovery [23], and cytotoxicity assays [24].

The process of fabricating microfluidic biochips on silicon wafers is described in [25]. In this process, the wafer is coated with a layer of SiO_2, which is used as the bulk insulation layer. Figure 1.4a shows electrodes fabricated on a silicon wafer. Two metal layers are deposited to form the interconnects. A cross-sectional view of a silicon-based biochip is shown in Fig. 1.4b [25].

The silicon-based microfluidic biochip in [25] can be used to generate and manipulate droplets that have volumes of 300 pl, and the actuation voltages required for dispensing and transporting a droplet are 11.4 V and 7.2 V, respectively.

1.1.3   Sensing Systems

The development of integrated biomedical analysis systems requires miniaturized, integrated, and robust sensing systems. Sensing systems on biochips can be recognized into three categories based on their working principle, i.e., droplet visualization monitoring system, on-chip capacitive sensor, and on-chip photodetector. More details about each of these systems are provided below.

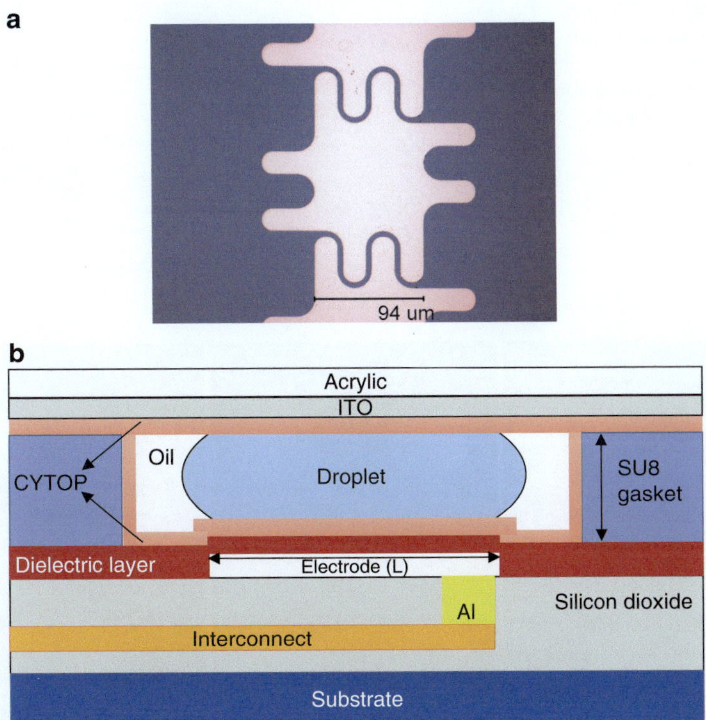

Fig. 1.4 (**a**) Electrodes of a silicon-based biochip [25]; (**b**) side view [25]

1.1.3.1 Droplet Visualization System

Cameras can be used to record the movements of multiple droplets on the biochip simultaneously. The setup of the droplet visualization system is shown in Fig. 1.5 [26]. By analyzing the images captured by the cameras, droplet dispensing, transportation, and mixing can be monitored. An example is shown in Fig. 1.6 [26]. One KCl droplet and one fluorescein droplet are mixed by repeating the splitting and merging operations. The time-lapsed images obtained from the top view show the procedure by which the fluorescein is diffused inside the droplet. Finally, the fluorescein will be distributed homogeneously inside the merged droplet.

When the bioassay is being conducted, the control software compares the images of the droplets in each intermediate step with the reference image of a fully-mixed droplet. In this way, the control software can determine the extent to which the mixing operation has approached completion [27]. In Sect. 2.2.1, we will show that the droplet visualization system can also be developed for error detection and for automatic tracking of droplets on biochips.

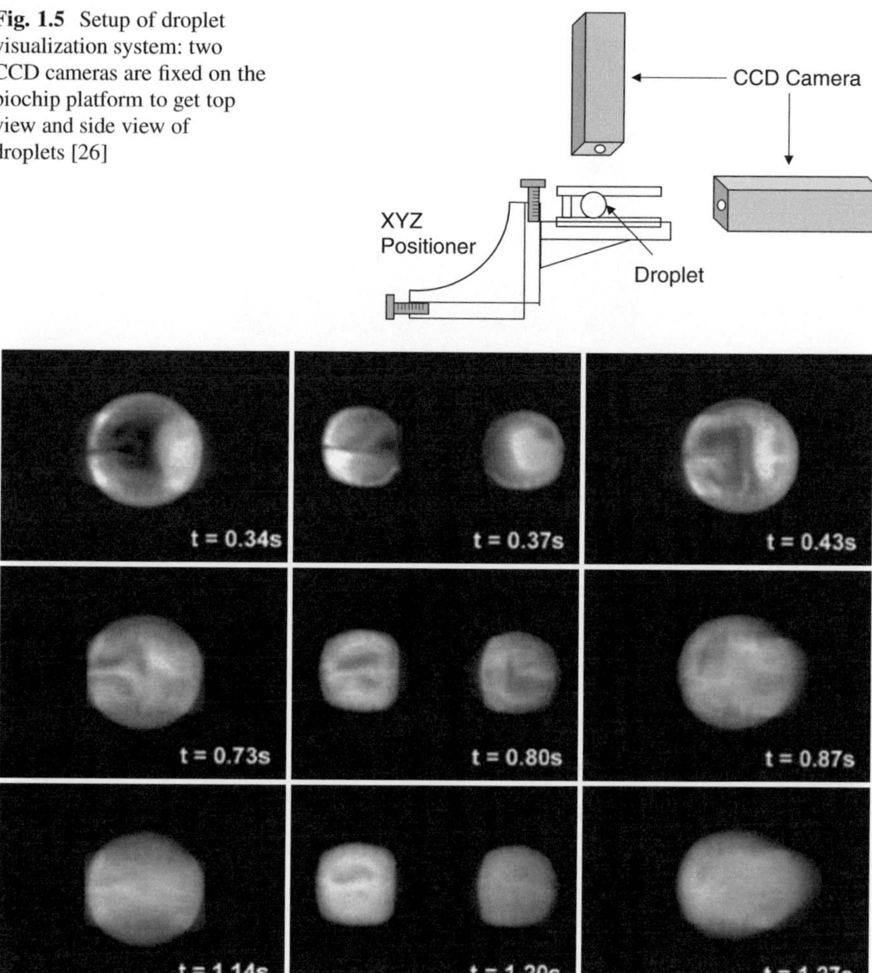

Fig. 1.5 Setup of droplet visualization system: two CCD cameras are fixed on the biochip platform to get top view and side view of droplets [26]

Fig. 1.6 Mixing of fluorescein and KCl droplets [27]

1.1.3.2   Capacitive Sensors

The upper and lower electrodes in each unit cell of the microfluidic biochip form a parallel-plate capacitor. The capacitance associated with the pair of electrodes can be measured by the resistor-capacitor (RC) circuit shown in Fig. 1.7 [28]. Since the droplet's permittivity is different from that of the filler medium of the biochip, the capacitance of a unit cell will change when a droplet is moved towards it. In this way, the movement of a droplet can be monitored.

Capacitive sensors can also be used to detect whether the droplet is contaminated by particles. As shown in Fig. 1.8 [29], a droplet may be contaminated by dust

Fig. 1.7 Circuit for capacitive sensing of biochip [28]

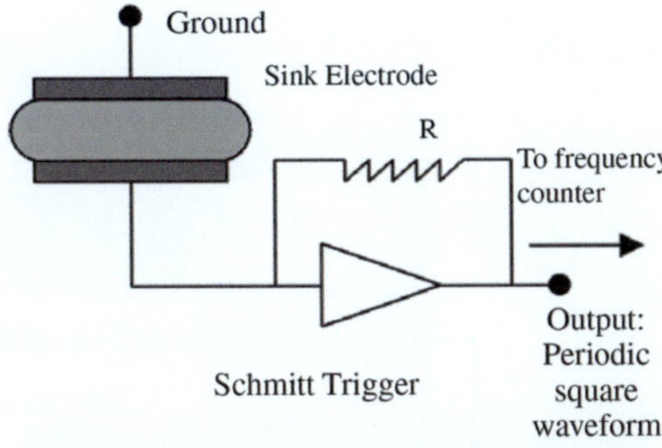

Fig. 1.8 A droplet with particle contamination [29]

particles or another small droplet that contains a different reagent. A contaminating particle changes the capacitance of the droplet. This difference can be detected and monitored by capacitive sensors. The minimum radius of a contaminating particle that can be detected by the capacitive sensor has been derived using theoretical calculations in [29].

By using the experimental setup shown in Fig. 1.9 [30], the relationship between the volume of droplet and the capacitance of a unit cell can be measured. During the measurement procedure, liquid is injected from the top of an electrode of the digital microfluidic biochip via a hole using a syringe pump. The flow rate of injected liquid is set as a constant, hence the volume of the droplet can be calculated based on the pumping time. By repeating the measurement procedures, the relationship about the volume of the droplet and the capacitance of electrode pair can be determined. Using this relationship and calibrated curve, the precise location and volume of each droplet can be estimated based on the feedback from capacitive sensors during bioassay execution.

1.1.3.3   Photodetectors and Optical Sensing

Photodetectors are widely used for sensing because they convert the intensity of fluorescence of droplet on chip into electrical signals [31–33]. These electrical signals can subsequently be applied as the feedback to the control software of the biochip.

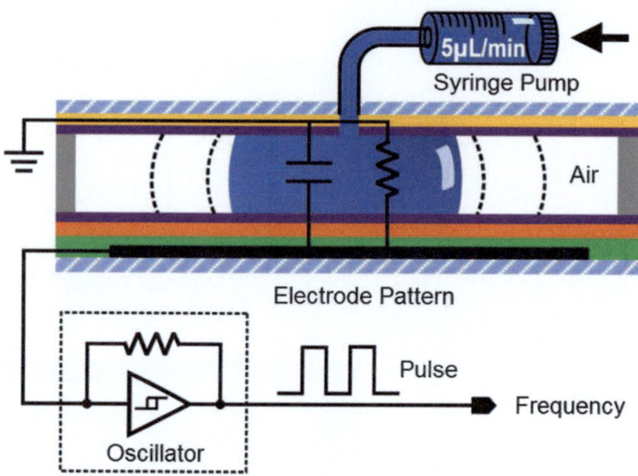

Fig. 1.9 Experimental setup for measuring the relationship between the volume of droplet and the capacitance of the unit cell [30]

1.1.4 Fault Models

Catastrophic and parametric faults may occur in digital microfluidic systems, which may affect the functionalities of the biochips [34]. Catastrophic faults may be induced by the following physical defects:

- *Dielectric breakdown:* When a high actuation voltage is applied, a electrical short between the droplet and the electrode may be created. As a result, no charge can be accumulated at the interface between the droplet and the electrode. As introduced in Sect. 1.1.1, the electrowetting mechanism is depended on the amount of energy stored in the capacitor formed by the electrode and the droplet. Hence, when dielectric breakdown occurs, droplet will stick to the electrode. Figure 1.10 shows a digital microfluidic biochip with dielectric breakdown [34, 35].
- *Short between the adjacent electrodes:* When an electrical short occurs between two adjacent electrodes, the two electrodes effectively form one electrode with longer length. As introduced in Sect. 1.1.1, in order to move a droplet, the droplet must contact electrodes that have different actuation voltages. If the droplet that resides on the longer electrode is not large enough to overlap the adjacent electrodes, the droplet cannot be moved [34].
- *Degradation of the electrode:* Degradation of electrodes is caused by their repeated use. This degradation effect can become catastrophic during the operation of the system. Figure 1.11 shows the degradation of an electrode [34].

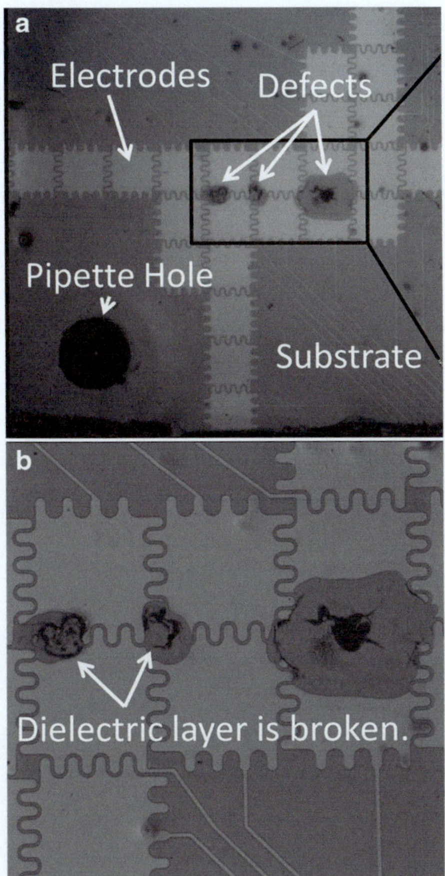

Fig. 1.10 (**a**) Digital microfluidic biochip with dielectric breakdown defects; (**b**) zoomed-in image of defects: When the dielectric layer is broken, the droplet(s) will make direct contact the metal electrode. When this happens, the droplet(s) stick to the electrode(s)

Fig. 1.11 Top view of the degradation of an electrode [34]

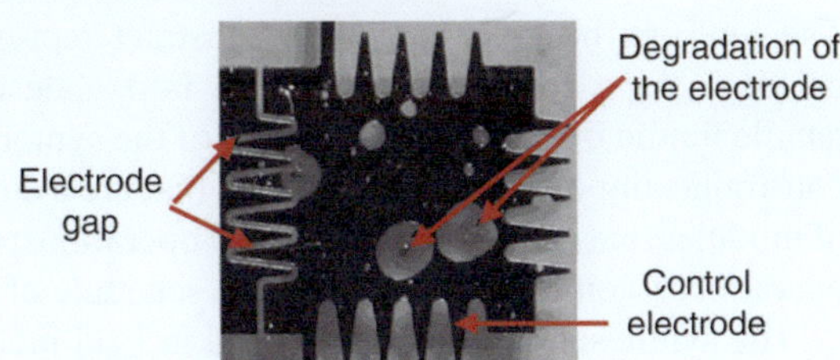

- *An opening in the metal connection between the electrode and the control source:* This defect results in the inability of charging the electrode while trying to drive the droplet.

Parametric faults may be caused by the following physical defects [34]:

- *Geometrical parameter deviation:* Due to the variations in the process, there may exist deviations in the thickness of the insulator layer, the dimensions of the electrodes, and the gap between the lower and upper plates. These deviations may affect the accuracy of the fluidic handling operations [34].
- *Insulator degradation:* In the insulation layer, ionization and slot discharge occur when the electrical field is high. The compounds formed during the discharges and the bombardment of ions can degrade the nearby insulating materials. The degradation of the insulation layer may gradually become apparent when the electrode performs fluidic operations. If the degradation of the insulator is left undetected, it may eventually affect the functionality of the biochip [34,36].
- *Particle contamination:* When bioassays are executed on a biochip, the droplet or the filler fluid on the biochip may be contaminated by dust particles or droplets of a foreign fluid [34].
- *Change in viscosity of droplet and filler medium:* When unexpected biochemical reactions occur on the biochip or the temperature varies in the biochip system, the viscosities of part of the droplets and the filler medium may change [34].

1.2   Computer-Aided Design and Optimization

In recent years, design automation methods for digital microfluidics have received significant attention [37–40]. This section discusses the research work that has been published in this area.

1.2.1   Design Flow for Digital Microfluidic Biochips

A unified synthesis tool for digital microfluidic-based biochips is described in [41]. The tool can be used to map an abstract representation of the desired bioassay protocol, e.g., a sequencing graph, into a design that can be implemented to handle fluidic operations. The inputs of the synthesis tool are the bioassay protocol, constraints imposed by the resources that are available on the biochip, and a library of modules that can implement fluidic operations; the outputs are a mapping of assay operations to on-chip resources and a schedule of the steps in the fluidic operations.

The synthesis procedure proposed in [41] includes architectural-level synthesis and geometry-level synthesis. Architectural-level synthesis can be viewed as the problem of scheduling fluidic handling operations and binding them to a given number of resources, whereas geometry-level synthesis addresses the placement of resources to satisfy area constraints. In [41], the parallel recombinative simulated annealing (PRSA) algorithm forms the core of the synthesis procedure that is used to obtain the optimized solution.

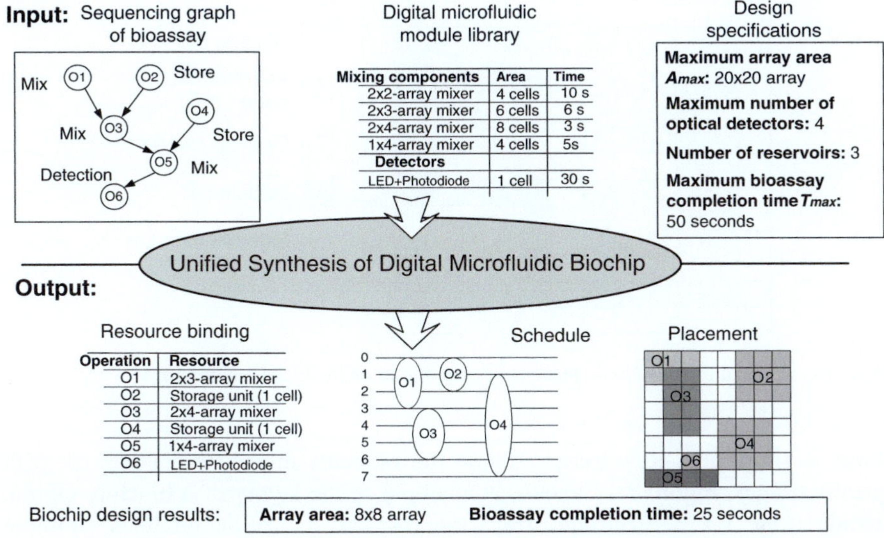

Fig. 1.12 The high-level synthesis procedure [41]

The PRSA-based algorithm for digital microfluidic biochips includes two parts, i.e., (1) the random generation of a large number of feasible synthesis configurations for the microfluidic biochips that satisfy all the resource and utilization constraints and dependency between operations and (2) the use of the PRSA algorithm to select an optimized synthesis configuration from these candidates. In this way, from a high-level specification of the bioassay, the tool can derive synthesis results that satisfy all the constraints related to on-chip resources and that minimize a given cost function for the given resource constraints. Figure 1.12 illustrates the steps in the flow of high-level synthesis.

A disadvantage of the synthesis procedure of [41] is that it fails to consider the routing of droplets. In some cases, all of the transportation paths for the droplets may be blocked by other modules on the biochip, and no feasible routing solution can be derived. To solve this problem, the authors of [42] proposed an improved PRSA-based synthesis algorithm that considers the time required for the transportation of a droplet as a criterion during design optimization. This approach can improve the routability of the synthesis results, even though it does not generate routing paths for the droplets or guarantee the existence of a routing solution.

Several algorithms for determining optimized droplet routing paths were proposed in [37, 40, 43, 44]. In [37], droplet routing was modeled as a motion-planning problem for multiple robots, and the A^* search technique was used to determine the shortest routing path.

A fast online synthesis method that enables real-time synthesis of biochips is proposed in [45]. In order to reduce the computational complexity and accelerate the online synthesis procedure, the geometry-level synthesis procedure introduced

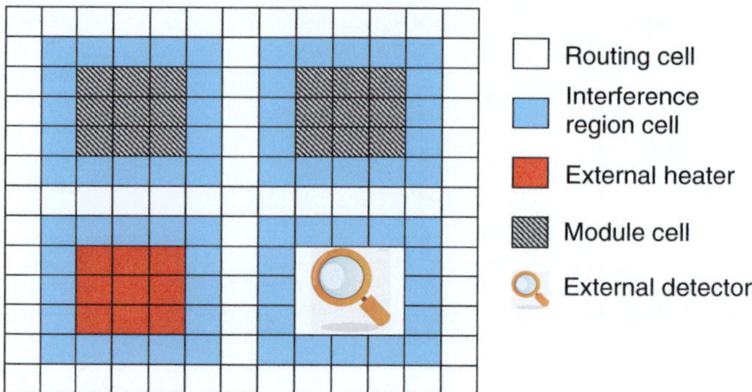

Fig. 1.13 Constraints for module placement on a digital microfluidic biochip [45]

above is simplified. Researchers divide the biochips into different regions with specific usages. Figure 1.13 shows an example of the layout of a biochip. On the biochip, four 3 × 3 electrode arrays are reserved as "basic modules". Fluidic operations, including the mixing, splitting, and storage of droplets, are performed by these modules. If there is an external detector or heater inside a basic module, that module also can perform detecting and heating operations.

The white cells in Fig. 1.13 represent the electrodes preserved for transporting droplets between modules and I/O ports. By using these dedicated routing cells, it is ensured that a valid path always exists between any source-destination pair. Surrounding each basic module, there are interference region (IR) electrodes, which are not used to perform any fluidic operation. Hence, the droplets inside the basic modules are isolated with the droplets in the routing paths, and interference-free droplet routes can be computed online easily.

When running the online synthesis procedure, the control software only determines the priority of the fluidic operations and then restricts the operations to the resources that are available on the biochip.

1.2.2 Testing Techniques

Many testing and diagnosis techniques have been proposed in recent years for use in detecting and locating catastrophic and parametric faults on biochips. In the test method proposed by [46], test stimuli droplets are dispensed into the biochip and then transported following the designed testing scheme.

In the testing scheme, the test stimuli droplets traverse every electrode on the biochip, after which they are moved to the droplet sink on the biochip. As discussed in Sect. 1.1.4, if an electrode has a catastrophic fault, droplets will stick to the electrode. Thus, if a digital microfluidic biochip has electrodes with catastrophic faults, test stimuli droplets will stick to them at intermediate points during the transportation. However, if all of the test stimuli droplets reach the droplet sink, there is no catastrophic fault on the biochip.

In order to minimize the total testing time for a given biochip, the test scheme must be optimized. Researchers in [46] found that the test planning problem can be formulated in terms of graph partitioning and the Hamiltonian path problem from graph theory. The two-dimensional microfluidic array is modeled as a direct graph, which is then divided into non-overlapping sub-graphs. Each sub-graph represents a sub-array on the biochip, and all of these sub-arrays are concurrently tested. In this way, the total time required for the test is reduced.

Researchers have also proposed an on-line error testing method that enables checking the correctness of the assay without hampering the normal operations in the bioassay [47]. Compared with the structural test introduced above, the online test procedure requires no extra test stimuli droplets, and it is finished when the bioassay is completed or sooner. The working principle of the proposed method is explained as follows.

First, "one time frame" was defined as the time required to perform a basic fluidic operation, such as dispensing a droplet, transporting a droplet to an adjacent electrode, merging two droplets in a mixer, and splitting a droplet. Hence, if a bioassay has n successive basic fluidic operations, the entire duration of the bioassay can be divided into n time frames, which are written as $f_1, f_2, \ldots, f_n$. Assume that the first droplet of the bioassay is dispensed into the biochip at the beginning of the bioassay and that the last droplet reaches the desired sink reservoir at the end of the bioassay.

At any time moment, the presence or absence of a droplet at a particular sink reservoir can be represented by 1 or 0, respectively. Assume that the bioassay is running on a biochip with s sink reservoirs, which are represented as $S_{R_1}, S_{R_2}, \ldots, S_{R_s}$. At time frame f_i, the expected "signature" of the bioassay can be written as an s-bit vector (r_i). If the droplet is supposed to reach the sink reservoir S_{R_j} at time f_i, then the j bit of r_i is 1, otherwise the j bit of r_i is 0.

Then, the expected signatures $r_1, r_2, \ldots, r_n$ can be derived based on the synthesis results of the bioassay. During the execution of the bioassay, the actual signature $r'_1, r'_2, \ldots, r'_n$ of the bioassay at each time frame can be captured by the integrated detectors on the bioassay.

By comparing the expected signatures and the actual signatures, researchers can tell whether there is a fault or error on the biochip. An error has occurred in the bioassay if there exists an integer $i (1 \leq i \leq n)$, such that $r_i \neq r'_i$. The assay is supposed to be executed without any error, and the outcome is accepted if for any integer $i (1 \leq i \leq n), r_i = r'_i$. In this way, the method can check the functionality of the biochip without any extra testing time or test stimuli droplets.

1.2.3　Error Recovery

For biomedical applications including clinical diagnostics, it is necessary to ensure the accuracy of the on-chip fluidic operations when the biochip is being used [48]. The accuracy of the fluid-handling operations can be monitored by examining various parameters, such as the volumes and concentrations of product droplets. If an error occurs during the execution of the bioassay, such as the volume of an intermediate product droplet exceeding the normal value, the assay outcomes can be incorrect, and the entire experiment must be repeated. Therefore, it is important to detect such errors as early as possible and repeat the corresponding fluid-handling operations to obtain correct outcomes from the bioassay.

In [48], the authors proposed a conceptual mechanism to monitor the intermediate products of the bioassay and implement error-recovery operations to minimize the influence of erroneous operations. A sequence of checkpoints is inserted into the initial protocol of the bioassay. At these checkpoints, sensors are used to check the qualities of the intermediate droplets. If the droplets fail to meet the quality requirements, the detector sends an interrupt to the control software, and some fluid-handling operations are repeated. In this way, error recovery can be implemented automatically.

1.2.4　Pin-Assignment Methods

The electrical I/O interface for digital microfluidics poses challenges. If each electrode were controlled by an independent pin and each pin were to have an input pad fabricated on the chip, the area required for the biochip would be extremely large, and the cost of fabrication would be high. Thus, several design optimization techniques have been proposed and analyzed in the literature [43, 49–51] with the goals of reducing the number of control pins and controlling the digital microfluidic array without significantly affecting concurrent droplet operations.

These methods reduce the number of pins in two different ways. One way is to reduce the number of pins by designing the biochip with some special structures. For example, in the n-bus-phase scheme, every nth electrode is connected to the same control pin. Another example is the cross-referencing biochip proposed in [52]. Figure 1.14 shows the cross-sectional views of the cross-referencing biochip. The rows and columns of electrodes in the biochip are fabricated on the top and bottom plates, respectively. A cell is activated when a row of electrodes are charged opposite polarities with a column of electrodes. When moving a droplet in the x-direction, the electrode columns on the top plate serve as the "ground electrode", and the electrode rows on the bottom plate serve as "driving electrodes" (Fig. 1.14a). Similarly, when moving a droplet in the y-direction, the electrode columns on the bottom plate serve as the "ground electrode", and the electrode rows on the top plate serve as "driving electrodes" (Fig. 1.14b).

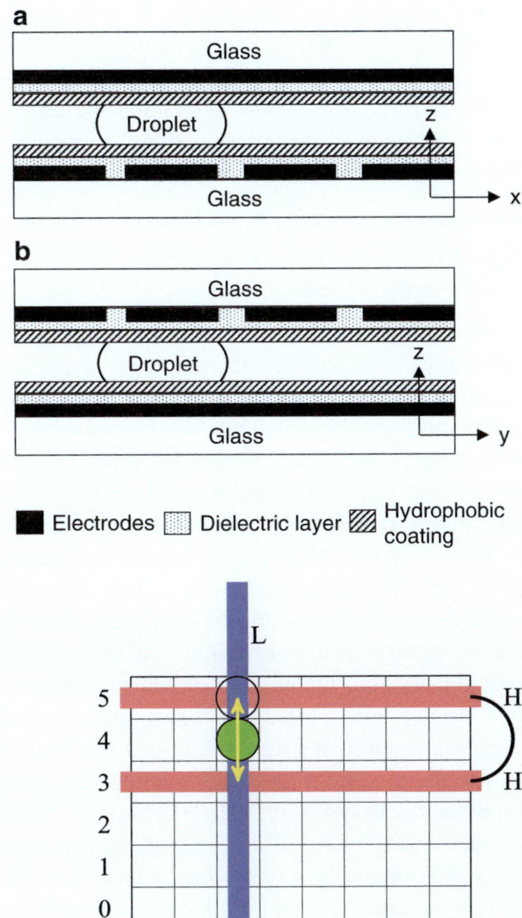

Fig. 1.14 Cross-section view of a cross-referencing biochip [45]

Fig. 1.15 An example of unwanted droplet splitting on the cross-referencing biochip [53]

From the above discussion, it is apparent that the cross-referencing design can greatly reduce the number of pins for a large-scale biochip. For a digital microfluidic biochip with $m \times n$ cells, a direct addressing biochip requires $m \times n$ pins, while a cross-referencing biochip only requires $(m + n)$ pins.

By connecting multiple columns/rows to the same control pin, the number of control pins on a cross-referencing can be further reduced [53]. However, this design may lead to unwanted movement or splitting of the droplets. Figure 1.15 [53] shows an example of this situation. On the biochip, row 3 and row 5 are controlled by a single pin. Assume that the droplet on cell (4, 2) (i.e., the cell in the fourth row and second column) is to be moved to cell (5, 2). Since rows 3 and 5 are controlled by the same pin, cell (2, 3) is also activated, and the droplet will be split. In fact, in this layout, no droplet can reach any of the electrodes on row 5.

For a sequence of electrodes $E = \{e_1, e_2, \ldots, e_n\}$, control pin mapping can be written as $Pin(E) = \{p_{\pi(1)}, p_{\pi(2)}, \ldots, p_{\pi(n)}\}$, where $p_{\pi(k)}$ is the pin that is connected to electrode e_k. $Pin(E)$ is defined as a control pin assignment sequence.

In [53], the researchers found that the unwanted movement/spliting of droplets can be avoided if the pin assignment sequence $P = \{p_1, p_2, \ldots, p_n\}$ satisfies the following constraints:

Constraint 1: $\forall p_i \in P$ *and* $\forall p_j \in P$ *where* $p_i = p_j$ *and* $i \neq j, \mid j - i \mid \geq 3$.
Constraint 2: $\forall p_i \in P$ *and* $\forall p_j \in P$, *if* $p_i = p_j$ *and* $i \neq j$, $\{p_{i-1}, p_{i+1}\} \cap$ $\{p_{j-1}, p_{j+1}\} = \emptyset$.

Based on the above constraints, the pin assignment configurations can be determined. It is also proven that an $m \times n$ pin-constrained cross-referencing biochip requires only $\sqrt{2}(\sqrt{m} + \sqrt{n})$ pins. In [53], a droplet routing method for cross-referencing biochips also was developed to maximize the degree of parallelism for fluidic operations. Even though the cross-referencing design of biochips can reduce the number of control pins, it results in increased fabrication cost since both the top and bottom plates of the cross-referencing biochip contain electrode rows/columns.

For a low-cost biochip that contains a discrete array of electrodes on the bottom plate, an alternative method to reduce the number of control pins is to connect multiple electrodes on the bottom plate directly to the same control pin. In [54], the researchers used a combinational logic circuit to generate control signals for the biochip, and thus the number of control pins was reduced significantly. The concept of a biochip with a logic control circuit is illustrated in the following example.

Let us assume that we have a 3×3 electrode array, and the electrodes are written as $E_1, E_2, \ldots, E_9$. Figure 1.16a shows the actuation signals that are to be applied on these electrodes, with "1", "0", and "*" representing "activated", "deactivated", or "don't care" status of an electrode at a specific time, respectively. Electrodes that possess compatible actuation sequences can be assigned to the same control pin. In this example, E_1 and E_9 have compatible actuation sequences, so they can be connected to the same control pin, as shown in Fig. 1.16b. This design is named as the "broadcast addressing scheme" of biochips [43].

However, if an actuation sequence of an electrode is to be generated as output signal from a logic function, the number of control pins for the electrode array can be reduced to two. Figure 1.16c presents the logic functions that correspond to the electrodes. Each electrode, E_i, is assigned a logic function $f(E_i)$, e.g., $f(E_1) = x_0$.

The example above presents the implementation of a logic circuit that can reduce the number of control pins significantly. In [54], the researchers further developed a scalable, heuristic approach to generate a control logic circuit for reducing the number of control pins.

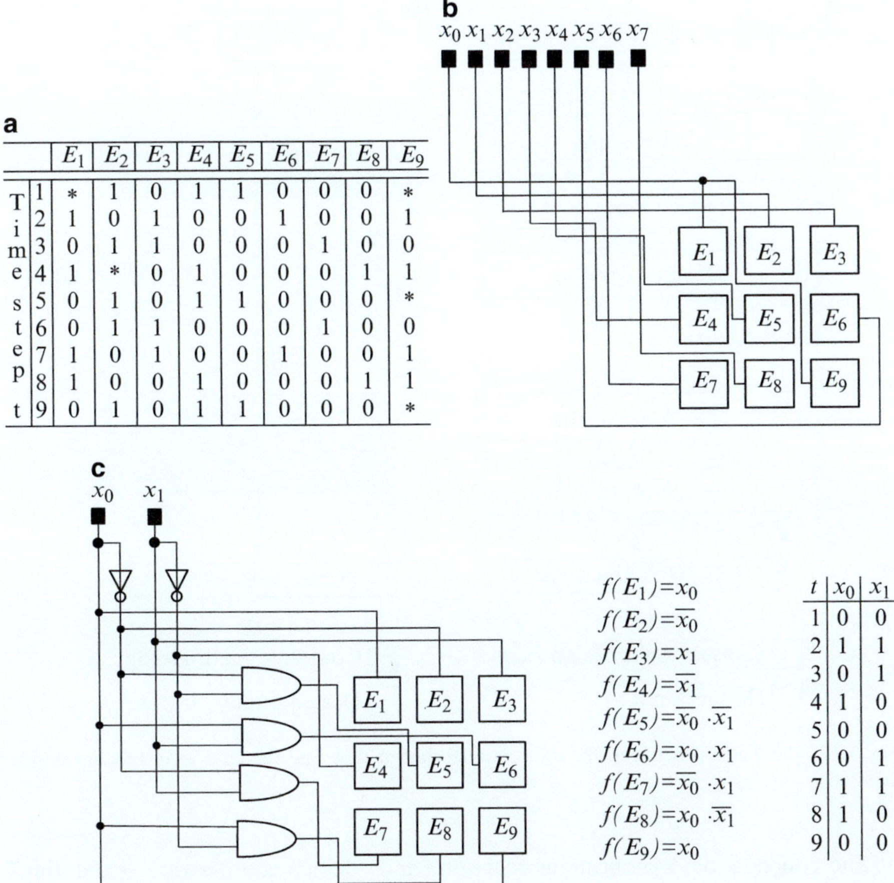

Fig. 1.16 (a) Actuation sequences to be applied on electrode; (b) broadcast-addressing biochip with 9 control pins; (c) pin-constrained biochip with control circuit

1.2.5 Chip-Level Design

The above discussion highlighted the design flow for digital microfluidic biochips. This design flow includes four stages, i.e., (1) high-level synthesis, (2) droplet routing, (3) derivation of the pin-assignment configuration, and (4) derivation of the wire-routing solution. Figure 1.17a illustrates the overall design flow [55]. "Fluidic-level synthesis" (which includes Stages 1 and 2) and "physical design" (which includes Stages 3 and 4) are optimized separately.

An integrated design flow of a biochip, which aims at filling the gap between fluidic-level synthesis and chip-level design, is proposed in [55]. The conventional and proposed design flows in [55] are compared in Fig. 1.17.

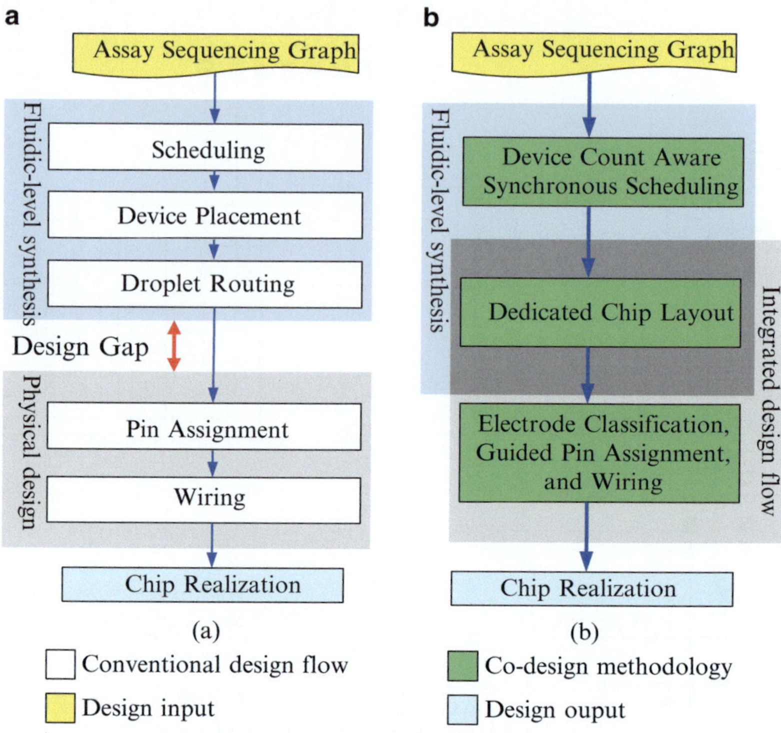

Fig. 1.17 Comparison between the conventional design flow and the chip-level biochip design flow [55]

The concepts of "synchronous reaction" and "device count-aware scheduling" are introduced in the synthesis stage. The reusability of modules can be improved, and the number of control pins required is minimized [55]. Next, the modules that correspond to these scheduled operations are placed on the biochip. Based on the device count-aware scheduling algorithm, the number of modules and the shapes of the modules are calculated. The placement of modules and the assignment of resources are determined to minimize the distances that the droplets must be transported. Electrodes are classified into three categories, i.e., bus, branch, and device electrodes. The pin-assignment methods and wire-routing strategies for these three categories are different. For example, the same types of devices that are used to implement synchronous operations share the same group of control pins. In this way, co-optimization for the synthesis result of the bioassay, pin-assignment configuration of the biochip, and metal wire-routing solution is achieved.

1.3   Outline of the Book

This book addresses a number of optimization problems related to cyberphysical microfluidic biochips. The reminder of the book is organized as follows.

Chapter 2 presents a transformative cyberphysical approach towards achieving closed-loop and sensor feedback-driven biochip operation under the program control. Section 2.1 presents the motivation for developing a "physical-aware" system reconfiguration technique that uses sensor data at intermediate checkpoints to dynamically reconfigure the biochip. Section 2.2 develops an algorithm for the measurement and tracking of droplets based on real-time imaging data from a charge-coupled device (CCD) camera. Section 2.3 introduces a reliability-driven error recovery strategy. Section 2.4 presents the parallel recombinative simulated annealing (PRSA)-based and greedy algorithms for reliability-driven synthesis. In Sect. 2.5, simulation results for three representative bioassays are discussed. Finally, conclusions are drawn in Sect. 2.6.

A hardware-assisted error-recovery method that relies on an error dictionary for rapid error recovery is presented in Chap. 3. The motivation for the hardware-assisted cyberphysical biochip is introduced in Sect. 3.1. The proposed algorithm for creation of the error dictionary is presented in Sect. 3.2. Section 3.3 introduces the generation of actuation matrices corresponding to synthesis solutions in the error dictionary. Section 3.4 illustrates that actuation matrices of bioassays are sparse. Based on this conclusion, Sect. 3.5 further discusses the procedures for compaction of the error dictionary. The implementation of dictionary-based error recovery on FPGA is introduced in Sect. 3.6. A fault simulation method with consideration of parameter variations in the fabrication process is discussed in Sect. 3.7. Simulation results are shown in Sect. 3.8, and conclusions are presented in Sect. 3.9.

Chapter 4 describes the synthesis algorithm for bioassays under completion-time uncertainties in fluidic operations. Section 4.1 discusses the drawbacks of previous uncertainty-oblivious methods of biochip design. Section 4.2 presents the design of microfluidic biochips with multiple clock frequencies. Section 4.3 introduces the framework of operation-dependency-aware synthesis. Based on results derived from the proposed synthesis algorithm, integrated on-line decision-making for droplet transportation path is presented in Sect. 4.4. Simulation results are presented in Sect. 4.5. Section 4.6 concludes the chapter.

Chapter 5 presents the optimization algorithms for PCR on a cyberphysical digital microfluidic biochip. The working principle of the PCR biochip is introduced in Sect. 5.1. Section 5.2 describes the statistical model for the amplification of DNA strands and the on-line decision making method during an actual experiment. Section 5.3 presents the device placement algorithm with the consideration of device interferences. Section 5.4 describes an application-specific reservoir allocation method. The consideration of droplet visibility during operation scheduling is discussed in Sect. 5.5. Simulation results for three widely used bioassays are presented in Sect. 5.6. Section 5.7 concludes the chapter.

The design procedure for a general-purpose microfluidic biochip is presented in Chap. 6. Section 6.1 describes previously published pin-assignment algorithms and their limitations. Section 6.2 presents an analysis of pin-actuation conflicts, and derives the necessary and sufficient conditions for control-pin sharing to ensure high flexibility in the concurrent movement of two droplets. Section 6.3 introduces an integer linear programming model for designing a pin-assignment with the smallest number of pins. Section 6.4 presents a graph-theoretic method to formulate an acceptance test for a pin-assignment configuration and a lower bound on the number of pins. A heuristic algorithm that generates a pin-assignment configuration for biochips is proposed in Sect. 6.4. Extension of the study from $1\times$ volume droplets to $2\times$ and even larger droplets is presented in Sect. 6.5. Section 6.6 presents the scheduling algorithm that can be applied to biochips with pin-constraints. Simulation results for commercial biochips and experimental prototypes are discussed in Sect. 6.7. Finally, conclusions are drawn in Sect. 6.8.

Based on the concepts discussed in Chaps. 2–4 and 6, the concept of pin-limited cyberphysical microfluidic biochip is proposed in Chap. 7. The structure and layout design of two-metal-layer biochips are introduced in Sect. 7.1. In Sect. 7.2 the wire-routing solution for general-purpose pin-limited biochips is proposed. The specific design flow for pin-limited cyberphysical biochips is discussed in Sect. 7.3. Results for several experimental bioassays are discussed in Sect. 7.4. Finally, conclusions are drawn in Sect. 7.5.

Chapter 8 summarizes the contributions of the book.

References

1. R. Fair, "Digital microfluidics: is a true lab-on-a-chip possible?", *Microfluidic and Nanofluidic*, vol. 3, pp. 245–281, 2007.
2. H. Becker, "Microfluidics: a technology coming of age", *Medical Device Technology*, vol. 19, 2008.
3. R. Fair, A. Khlystov, T. Tailor, V. Ivanov, R. Evans, P. Griffin, V. Srinivasan, V. Pamula, M. Pollack, and J. Zhou, "Chemical and biological applications of digital-microfluidic devices", *IEEE Design & Test of Computers*, vol. 24, pp. 10–24, 2007.
4. F. Su, K. Chakrabarty and R. B. Fair, "Microfluidics-based biochips: technology issues, implementation platforms, and design automation challenges", *IEEE Transactions on Computer-Aided Design of Integrated Circuits & Systems*, vol. 25, pp. 211–223, February 2006.
5. K. Chakrabarty, R. Fair and J. Zeng, "Design tools for digital microfluidic biochips: Towards functional diversification and more than Moore", *IEEE Trans. CAD*, vol. 29, pp. 1001–1017, 2010.
6. H.-C. Chang, and L. Yeo, *Electrokinetically Driven Microfluidics and Nanofluidics"*, New York, NY: Cambridge University Press, 2009.
7. http://www.cytonix.com/fluid
8. (Fluidigm Corporation) http://www.fluidigm.com/
9. P. Marcoux, M. Dupoy, R. Mathey, A. Novelli-Rousseau, V. Heran, S. Moralesa, F. Riveraa, P. Joly, J. Moy, and F. Mallard, "Micro-confinement of bacteria into w/o emulsion droplets for rapid detection and enumeration", *Colloids and surfaces A :Physicochemical and engineering aspects*, vol. 377, no 1–3, pp. 54–62, 2011.

10. Silicon Biosystems. www.siliconbiosystems.com.
11. B. Kirby, *Micro- and Nanoscale Fluid Mechanics: Transport in Microfluidic Devices*, New York, NY: Cambridge University Press, 2010.
12. R. Tadmor, "Line energy and the relation between advancing, receding and Young contact angles", *Langmuir*, Volumne 20, Issue 18, pp. 7659–7664, 2004.
13. P.-G. Gennes, F. Brochard-Wyart, and D. Quere, *Capillary and Wetting Phenomena - Drops, Bubbles, Pearls, Waves*, New York, NY: Springer Press, 2004.
14. D. Woodruff, *The Chemical Physics of Solid Surfaces*, Amsterdam, Netherlands: Elsevier, 2002.
15. F. Saeki, J. Baum, H. Moon, J. Yoon, C.-J. Kim, and R. Garrell, Polym, "Electrowetting on dielectrics (EWOD): reducing voltage requirements for microfluidics mater", *Sci. Eng*, Issue 85, pp. 12–13, 2001.
16. J. Reed and K. Guthe, *College Physics*, Charleston, SC: Nabu Press, 2010
17. M. Pollack, R. Fair, and A. Shenderov, "Electrowetting-based actuation of liquid droplets for microfluidic applications", *Appl. Phys Lett*, vol. 77, pp. 1725–1727, 2000.
18. J. Lee, H. Moon J. Fowler, C.-J Kim and T. Schoellhammer, "Addressable micro liquid handling by electric control of surface tension", *Proc. of 2001 IEEE 14th International Conference on MEMS*, pp. 499–502, 2001.
19. S.-K. Cho et. al., "Towards digital microfluidic circuits: creating, transporting, cutting and merging liquid droplets by electrowetting-based actuation", *IEEE International Conference on Micro Electro Mechanical Systems*, pp. 454–461, 2002.
20. R. Fair, A. Khlystov, V. Srinivasan, V. Pamula, and K. Weaver, "Integrated chemical/biochemical sample collection, pre-concentration, and analysis on a digital microfluidic lab-on-a-chip platform", *Proceedings of SPIE*, volume 5591, Issue 8, page 113–124, 2004.
21. Advanced Liquid Logic, http://www.liquid-logic.com.
22. T. Xu and K. Chakrabarty, "Functional testing of digital microfluidic biochips", *Proc. IEEE International Test Conference*, pp. 1–10, 2007.
23. H. Ren, V. Srinivasan, and R. Fair, "Design and testing of an interpolating mixing architecture for electrowetting-based droplet-on-chip chemical dilution", *International Conference on Solid-State Sensors, Actuators and Microsystems*, Volume 1, pp. 619–622, 2003.
24. I. Nad, H. Yang, P. Park, and A. Wheeler, "Digital microfluidics for cell-based assays", *Lab Chip*, Volume 8, Issue 4, pp. 519–526, 2008.
25. Y.-Y. Lin, R. Evans, E. Welch, B.-N. Hsu, A. Madison, and R. Fair, "Low voltage electrowetting-on-dielectric platform using multi-layer insulators", *Sensors and Actuators, B: Chemical*, vol. 105, pp. 465–470, 2010.
26. V. Pamula, P. Paik, J. Venkatraman, M. Pollack, and R. Fair, "Microfluidic electrowetting-based droplet mixing", *IEEE MEMS Conference Proceedings*, pp. 8–10, 2001.
27. P. Paik, V. Pamula, and R. Fair, "Rapid droplet mixers for digital microfluidic systems", *Lab on a Chip*, vol. 3, pp. 253–259, 2003.
28. M. Pollack, *Electrowetting-based Microactuation of Droplets for Digital Microfluidics*, PhD Thesis, Duke University, Durham, NC, 2001.
29. T.-Y. Ho, K. Chakrabarty and P. Pop, "Digital microfluidic biochips: Recent research and emerging challenges", *Proc. IEEE CODES+ISSS*, 2011.
30. J. Gao, X. Liu, T. Chen, P.-I. Mak, Y. Du, M. Vai, B. Lin, and R. Martins, "An intelligent digital microfluidic system with fuzzy enhanced feedback for multi-droplet manipulation", *Lab on a Chip*, Issue 13, volume 13, 2013.
31. H. Dutton, *Understanding Optical Communications*, Upper Saddle River, New Jersey: Prentice Hall Press, 1998.
32. N. Jokerst, L. Luan, S. Palit, M. Royal, S. Dhar, M. Brooke, and T. Tyler II, "Progress in chip-scale photonic sensing", *IEEE Trans. Biomedical Circuits and Sys.*, vol. 3, pp. 202–211, 2009.
33. R. Evans et. al., "Optical detection heterogeneously integrated with a coplanar digital microfluidic lab-on-a-chip platform", *Proc. IEEE Sensors Conf.*, pp. 423–426, Oct. 2007.

34. F. Su, S. Ozev and K. Chakrabarty, "Ensuring the operational health of droplet-based microelectrofluidic biosensor systems", *IEEE Sensors*, vol. 5, pp. 763–773, August 2005.
35. K. Hu, B.-N. Hsu, A. Madison, K. Chakrabarty and R. Fair, "Fault detection, real-time error recovery, and experimental demonstration for digital microfluidic biochips", *Proc. IEEE/ACM Design, Automation and Test in Europe (DATE) Conference*, pp. 559–564, 2013.
36. D. Tommasini, "Dielectric insulation and high-voltage issues", arXiv:1104.0802v1, 2011.
37. K. Bohringer, "Modeling and controlling parallel tasks in droplet-based microfluidic systems", *IEEE Transactions on Computer-Aided Design of Integrated Circuits & Systems*, vol. 2, pp. 329–339, 2006.
38. P.-H. Yuh, C.-L. Yang, and Y.-W. Chang, "Placement of digital microfluidic biochips using the T-tree formulation", *Proc. IEEE/ACM Design Automation Conference*, pp. 931–934, 2006.
39. M. Cho and D. Z. Pan, "A high-performance droplet router for digital microfluidic biochips", *Proc. ACM International Symposium on Physical Design*, pp. 1714–1724, 2008.
40. T.-W. Huang, C.-H. Lin, and T.-Y. Ho, "A contamination aware droplet routing algorithm for the synthesis of digital microfluidic biochips", *IEEE Transactions on Computer-Aided Design of Integrated Circuits and Systems*, vol. 29, no. 11, pp. 1682–1695, 2010.
41. K. Chakrabarty and F. Su, *Digital Microfluidic Biochips: Synthesis, Testing, and Reconfiguration Techniques*, Boca Raton, FL: CRC Press, 2006.
42. T. Xu and K. Chakrabarty, "Integrated droplet routing in the synthesis of microfluidic biochips", *Proc. IEEE/ACM Design Automation Conference*, pp. 948–953, 2007.
43. T. Xu and K. Chakrabarty, "Broadcast electrode-addressing for pin-constrained multifunctional digital microfluidic biochips", *Proc. IEEE/ACM Design Automation Conference*, pp. 173–178, 2008.
44. Y. Zhao and K. Chakrabarty, "Simultaneous optimization of droplet routing and control-pin mapping to electrodes in digital microfluidic biochips", *IEEE Transactions on Computer-Aided Design of Integrated Circuits and Systems*, vol. 31, pp. 242–254, February 2012.
45. D. Grissom and P. Brisk, "Fast online synthesis of generally programmable digital microfluidic biochips", *Proc. CODES+ISSS*, pp. 413–422, 2012.
46. F. Su, S. Ozev, K. Chakrabarty, "Test planning and test resource optimization for droplet-based microfluidic systems", *Journal of Electronic Testing: Theory and Applications*, Volume 22 Issue 2, pp. 199–210, April 2006.
47. D. Mitra, S. Ghoshal, H. Rahaman, K. Chakrabarty, and B. Bhattacharya, "On-line error detection in digital microfluidic biochips", *Proc. IEEE Asian Test Symposium*, pp. 332–337, 2012.
48. Y. Zhao, T. Xu, and K. Chakrabarty, "Integrated control-path design and error recovery in digital microfluidic lab-on-chip", *ACM JETC*, Vol. 6, No. 3, Article 11, 2010.
49. V. Srinivasan, V. Pamula, and R. Fair, "An integrated digital microfluidic lab-on-a-chip for clinical diagnostics on human physiological fluids", *Lab on a Chip*, vol. 4, pp. 310–315, 2004.
50. Z. Xiao and E. Young, "CrossRouter: A droplet router for cross-referencing digital microfluidic biochips", *IEEE/ACM Asia South Pacific Design Automation Conference*, pp. 269–274, 2010.
51. C.-Y. Lin and Y.-W. Chang, "Cross-contamination aware design methodology for pin-constrained digital microfluidic biochips", *IEEE Transactions on Computer-Aided Design of Integrated Circuits and Systems*, vol. 30, Issue 6, pp. 817–828, 2011.
52. S.-K. Fan, C. Hashi, C.-J. Kim, "Manipulation of multiple droplets on $N \times M$ grid by cross-reference EWOD driving scheme and pressure-contact packaging", *IEEE International Conference on Micro Electro Mechanical Systems*, pp. 694–697, 2003.
53. H.C. Yeung and F.Y. Young, "General purpose cross-referencing Microfluidic Biochip with reduced pin-count", *Asia and South Pacific Design Automation Conference*, pp. 238–243, 2014.
54. T. A. Dinh, S. Yamashita, and T.-Y. Ho, "A logic integrated optimal pin-count design for digital microfluidic biochips", *Proceedings of the conference on Design, Automation & Test in Europe*, pp. 1–6, 2014.
55. T.-W. Huang, J.-W. Chang, and T.-Y. Ho, "Integrated fluidic-chip co-design methodology for digital microfluidic biochips", *Proceedings of ACM International Symposium on Physical Design*, pp. 49–56, 2012.

Chapter 2
Error-Recovery in Cyberphysical Biochips

In this chapter, through exploiting recent advances in the integration of sensing system in a digital microfluidics biochip, we present a "physical-aware" system reconfiguration technique that uses sensor data at intermediate checkpoints to reconfigure the biochip dynamically. A cyberphysical re-synthesis technique is used to recompute electrode-actuation sequences, thereby deriving new results for module placement, droplet routing pathways, and operation schedules, with minimum impact on the time-to-response.

The key contributions of this chapter are as follows:

- A charge-coupled device (CCD)-based sensing system for digital microfluidic biochips (Sect. 2.2).
- An imaging algorithm for the measurement and tracking of droplets based on real-time data from a CCD camera (Sect. 2.2).
- A reliability-driven error-recovery strategy (Sect. 2.3).
- Parallel recombinative simulated annealing (PRSA)-based and greedy algorithms for reliability-driven synthesis (Sect. 2.4).
- Simulation results for three representative bioassays (Sect. 2.5).

2.1 Motivation and Related Prior Work

The ease of reconfigurability and software-based control in digital microfluidics has inspired research on various aspects of automated chip design and chip applications. A number of techniques have been published for architectural-level synthesis [1], module placement, and droplet routing [2–4]. However, these techniques ignore practical realities or domain-specific constraints that arise from attempting to carry out biochemical reactions and microfluidic operations on an electronic chip.

© Springer International Publishing Switzerland 2015

Y. Luo et al., *Hardware/Software Co-Design and Optimization for Cyberphysical Integration in Digital Microfluidic Biochips*, DOI 10.1007/978-3-319-09006-1_2

Due to the complex and randomness component interactions that are ubiquitous in biological/chemical processes, predictive modeling and accuracy control become difficult [5, 6].

In addition to manufacturing defects and imperfections, various faults may also arise during bioassay execution. For example, DNA fouling may lead to malfunction of multiple electrodes in the biochip and excessive actuation voltage applied to an electrode may lead to breakdown of electrodes and charge trapping [7–9]. These faults are hard to detect a priori, but they occur during bioassays in certain situations [9]. Yet, despite such inherent variability, many biomedical experiments, such as clinical diagnostics and drug development, require fluid-handling operations that are highly accurate and precise. Each step in the protocol of a biochemical experiment has an "acceptance range" for the volume and concentration of droplets. For example, in the preparation of samples of plasmid DNA, the pH of the solution must be less than 8.0 to avoid a significant reduction in the efficiency of the lysozyme [10]. If an unexpected error occurs during the experiment or the requirements of the bioassay protocol are violated, the outcome of the entire experiment are incorrect. When this occurs, all the steps of the experiment must be repeated to correct the error [11, 12]. Repetition of experiments leads to wastage of expensive reagents and hard-to-obtain samples.

The repetitive execution of on-chip laboratory experiments results in the following problems: (i) an increase in the time-to-result for a bioassay, which is detrimental to real-time detection and rapid response; (ii) wastage of samples that are difficult to obtain or prepare, as well as the wastage of expensive reagents.

Therefore, it is necessary to develop techniques for monitoring assay outcomes at intermediate stages and design an efficient error-recovery mechanism. Error recovery in digital microfluidics has received relatively little attention in the literature. The only reported work is [11], which proposed intermediate stage monitoring and rollback error-recovery for a microfluidic biochip. In [11], sensing system is used to verify the correctness of immediate product droplets at various steps in the on-chip experiment. When an error is detected at a sensor, i.e., the volume or concentration of the droplet is below or above the acceptable calibrated range, the corresponding droplet is discarded. If the outputs of an operation fail to meet the quality requirements that are derived based on sensor calibration, the operation will be re-executed. New product droplets will be generated to replace the unqualified droplet.

Figure 2.1 shows an example of rollback error-recovery. The initial sequencing graph of a bioassay is shown in Fig. 2.1a. Here we assume that the outputs of each dispensing, mixing and splitting operation are evaluated by a sensor. When an error occurs at operation 9, the system will re-execute the corresponding dispensing and mixing operations. Figure 2.1b shows the new sequencing graph for error-recovery, where operations 12, 13, and 14 are added to generate new product droplets. In the absence of "physical-aware" control software, the error-recovery method in [11] suffers from following drawbacks:

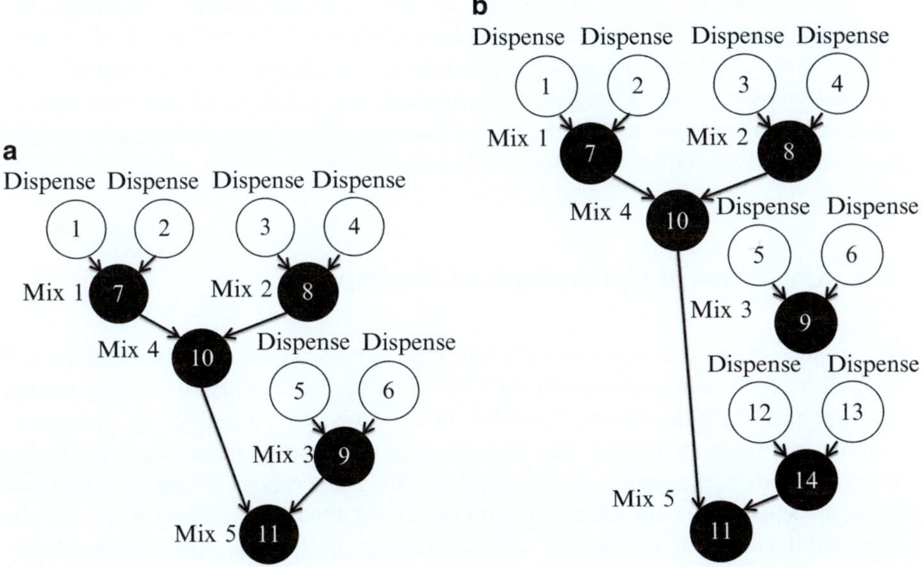

Fig. 2.1 (**a**) Initial sequencing graph; (**b**) operations 12, 13, and 14 are added for error-recovery

1. The first drawback is the over-simplification of fault detection and the associated assumptions. It is impractical to use a uniform "expected value" for the calibration of each detection operation. Note that the concentration of intermediate product droplets vary in a dynamic manner at various stages during bioassay execution. Hence the calibration needs to be repeated and carried out dynamically as well.

2. In [11], all recovery operations are carried out in a stand-alone manner. All other ongoing bioassay-related fluidic operations are interrupted when an error is detected. The potential long waiting times introduced by recovery operations will lead to sample degradation and erroneous assay outcomes [14]. Some operations, such as colorimetric enzyme-kinetic reactions, require precise durations as specified by the reaction protocol, and they cannot be elongated without introducing unpredictability in the experiment outcome [15].

3. The error-recovery approach in [11] cannot handle situations when multiple errors occur during a bioassay. For example, [11] assumes that all error-recovery operations will be executed successfully and it does not consider the likelihood that errors can also occur during recovery.

4. The error-recovery strategy in [11] does not consider reliability issues. Errors such as the generation of droplets with abnormal volumes are usually caused by the accumulation of charge on the surface of certain electrodes [7, 8]. If the use of such electrodes is continued, it is likely that they will introduce more errors [7, 8]. Thus, in order to ensure the reliability of biochips, we must minimize the utilization of these electrodes.

To overcome the above drawbacks, we take a transformative "cyberphysical" approach towards achieving sensor feedback-driven and closed-loop biochip operations under software control. By exploiting recent advances in the integration of sensing system in a digital microfluidics biochip [16], we present a "physical-aware" system reconfiguration technique that uses sensor data at intermediate checkpoints to dynamically reconfigure the biochip [17].

2.2 Overview of Cyberphysical Biochips

In this section, we introduce the cyberphysical system on microfluidic biochips and as well as each component of the system. With the availability of sensing system for biochips, "physical-aware" control software becomes feasible. By "physical-aware", we refer to the fact that the software can receive information about the outcome (error-free/erroneous) of fluid-handling operations based on feedback from the sensing system. Depending on sensor feedback, the control software will appropriately reconfigure the microfluidic biochip. In this way, the various steps in the bioassay are executed based upon real-time sensing of intermediate results.

Figure 2.2 depicts each component of a cyberphysical system on the microfluidic platform. The control software sends a control signal to the microfluidic biochip, and the on-chip sensing system monitors the outcomes of fluidic operations. The outcomes are compared with the "expected values", i.e., the pre-determined thresholds. If the results of the comparison indicate that an error has occurred, the control software receives a "repeat request", and the corresponding operation in which the error occurred can be executed again. In this way, the error will be corrected.

2.2.1 Sensing Systems

Two sensing schemes can potentially be used in the cyberphysical system on digital microfluidic biochips.

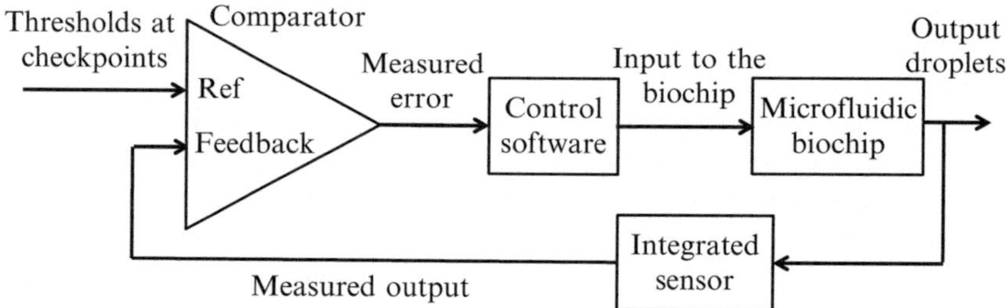

Fig. 2.2 The schematic workflow of the cyberphysical digital microfluidic system

The first sensing scheme is CCD camera-based. As described in Chap. 1, CCD cameras can be used in experiments to view the top sides of droplets simultaneously.

Based on images captured by the CCD camera, droplets can be automatically located by the control software. The procedure for automatical search of the droplets can be described as a "template matching" problem. Here a pattern can be represented as the image of a "typical" droplet. During the matching process, we move the template image to all possible positions in the image of the entire array and crop a sub-image that has the same size as the template image. Then the control software computes the correlation index, which measures the similarity between the template and the "cropped image". The correlation factor is calculated on a pixel-by-pixel basis, and this process is shown in Fig. 2.3a.

In the control software, all images are stored in grayscale form. These grayscale images can be encoded as matrices or vectors. Suppose the template image is represented in a 1-D array: $\mathbf{x} = (x_1, x_2, \ldots x_N)$. Here x_i represents the gravel level of a pixel and N is the total number of pixels in the template image. Similarly, the cropped sub-image to be compared with the template image can be written as $\mathbf{y} = (y_1, y_2, \ldots, y_N)$. Thus the correlation factor between these two images is defined as:

$$cor = \frac{\sum\limits_{i=1}^{N} (x_i - \bar{x}) \cdot (y_i - \bar{y})}{\sqrt{\sum\limits_{i=1}^{N} (x_i - \bar{x})^2 \cdot \sum\limits_{i=1}^{N} (y_i - \bar{y})^2}},$$

where $\bar{x}$ and $\bar{y}$ are the average gray level in the template image and cropped sub-image, respectively. The range of correlation factor cor is a real number between -1 and $+1$. According to the definition of correlation, a correlation factor with larger absolute value represents a stronger relationship between two images.

After deriving the correlation factors for all possible locations in the image of the complete biochip, we obtain the correlation map between the template and the original input image. Suppose there are κ droplets on the biochip. The locations of droplets can be determined by searching for the largest κ correlation factors in the correlation map. An example is shown in Fig. 2.3b, c [18].

Figure 2.3b shows the original input image of the whole chip and the pattern image, and Fig. 2.3c is the correlation map, where the best matching locations, i.e. the coordinations of droplets derived by the control software are (77, 107), (77, 147) and (76, 208). Thus the control software automatically locates the droplets. According to the image, the sizes and colors of droplets can be further analyzed. In this manner, the volumes and concentrations of droplets can be acquired after processing the image taken by the CCD camera.

Instead of searching for droplets in the complete image, we use imaging techniques to check whether the droplets have been moved to the expected positions. This procedure is implemented using the following steps:

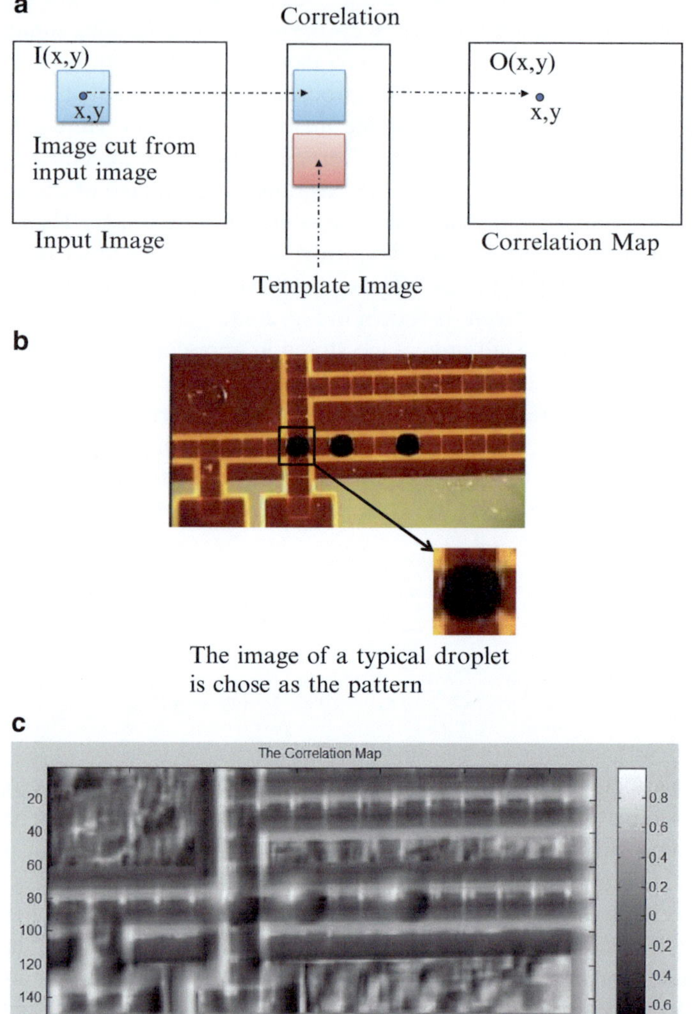

Fig. 2.3 (**a**) The matching process moves the template image to all possible positions in a larger source image and computes a numerical index that indicates how well the template matches the sub-image in that position; (**b**) the image of the whole biochip [18] and the pattern we selected; (**c**) the correlation map between image of the whole array and the pattern. The positions of droplets can be determined by finding κ maximum elements (κ is the number of droplets on the chip) in the correlation map

First, some calibrations are made before the experiment. We choose a large number of sub-images with (or without) droplets, and calculate their correlation with the template. Based on the calculation results, we search an appropriate threshold for the correlation index (C_{th}): if the correlation is larger than C_{th}, we conclude that there is a droplet in the cropped sub-image; otherwise, there is no droplet in the sub-image. When the bioassay is running, we only need to crop the sub-images near the expected positions of droplets, and calculate their corresponding correlation indices to determine the absence/presence of droplets.

The advantages of the CCD camera-based sensing system are: (i) the identification of the precise locations of the errors, and (ii) the detection of errors immediately after they occur. One disadvantage of this system is that extra instruments, such as CCD cameras, are required to observe the cyberphysical system.

The second sensing scheme is based on integrated optical detectors, as proposed in [15, 16]. By examining the concentration of the product in the droplets through fluorescence, the quality of an intermediate product in a digital microfluidic biochip can be determined [15, 16].

When a fluorophore tag is attached to a droplet, different product concentrations lead to the emission of light with different spectrum (i.e., different colors). This difference in color can be detected by optical sensors that convert the received light into electrical current or a voltage signal [16]. In recent work, integrated photodetectors have been introduced on the microfluidic array [15, 16]. For example, in [15], an optical detection system was integrated with the digital microfluidic array. It consists of a light-emitting diode (LED) and a photodiode which functions as light-to-voltage converter. The concentration of products can be calculated according to the output voltage of the photodiode. Another example, thin film InGaAs photodetectors can be bonded onto a glass platform, coated with Teflon AF, and then integrated into the digital microfluidic system. A coplanar digital microfluidic chip with the integrated InGaAs photodetector is shown in Fig. 2.4 [15].

Even though no instruments with large footprint and precise alignment are required in this method, the integrated optical detector-based sensing system has a drawback that it cannot precisely locate the electrode where an error has occurred. For example, when an output droplet of an operation is sent to the detector and it fails to meet the requirement of the bioassay, we know that an error occurred during the mixing operation, but we cannot locate the precise time and the position where it occurred. The comparison between CCD camera-based sensing scheme and detector-based sensing scheme can be found in Table 2.1.

Fig. 2.4 Coplanar digital microfluidic chip with integrated thin film photodetector [15]

Table 2.1 Comparison between CCD camera-based sensing scheme and detector-based sensing scheme

	CCD camera-based scheme	Detector-based scheme
Accuracy of locating electrodes with defect	High	Low
Response time	Error recovery can be triggered immediately when an error occurred	Error recovery can only of the faulty operation be triggered at the end
Application for photosensitive samples/reagents	Cannot be used	Can be used
Cost	The price of the CCD camera is around \$3,000	The cost of fluorescent labels is around \$30/μ mol and the Laser is around \$200

2.2.2 *"Physical-Aware" Software*

The availability of on-chip sensors provides digital microfluidic biochips with the capability of using sensor data at intermediate checkpoints to detect errors, thereby minimizing the impact of errors that occur during bioassay execution. The work in [11] proposed intermediate stage monitoring and rollback error-recovery for a microfluidic biochip. The key idea in this work is using the sensing system on-chip to verify the correctness of output droplets at various steps in the on-chip experiment. In this approach, error-recovery is carried out as follows. When an error is detected at a checkpoint, operations whose outputs failed to meet the quality requirements based on sensor calibration are re-executed to recover from the error. The unused intermediate product droplets must be stored in specially designated locations of the chips to facilitate recovery. Additional droplets of samples and reagents must also be dispensed from reservoirs for error-recovery.

Based on the error-recovery mechanism proposed in [11], the strategy for reliability-driven error-recovery will be introduced in Sect. 2.3, and the algorithm for dynamic re-synthesis of error-recovery will be described in Sect. 2.4.

2.2.3 *Interfaces Between Biochip and Control Software*

We next describe the cyberphysical coupling between the control software and the hardware of the microfluidic platform. There are two interfaces needed for cyberphysical coupling. The first interface converts the output signals from the sensing system to the desktop computer, thus the control software can interpret the feedback signals. The second interface converts the output data generated by the

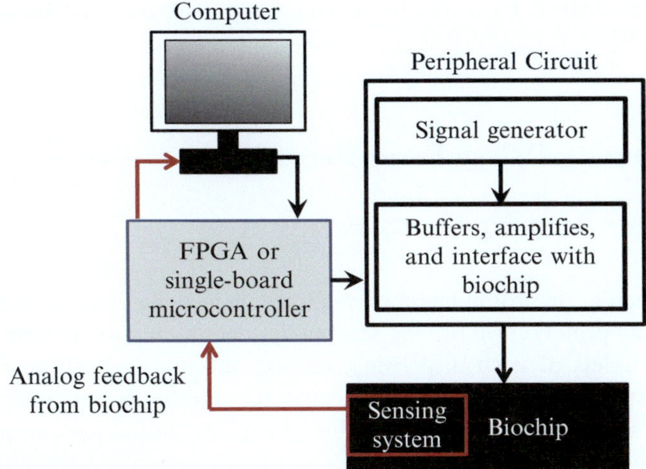

Fig. 2.5 The schematic of the cyberphysical digital microfluidic system. Software running on the computer and the biochip are coupled by a peripheral circuit and a field-programmable gate array (or a single-board microcontroller)

control software to voltage signals that can be directly applied to the electrodes of the biochip.

Figure 2.5 shows a schematic representation of the cyberphysical system. The system consists of a computer, a peripheral circuit, a single-board microcontroller, and the biochip. During the execution of a bioassay, the software running on the computer sends control signals to the biochip. The software at the same time receives feedback from the sensing system on the biochip through the single-board microcontroller. The control software recomputes the schedule of fluidic operations, module placement, and droplet pathways depending on sensor feedback. This process is referred to as reconfiguration and the key idea is to use an on-chip sensing system for concurrent monitoring of experiments. When an error is detected by on-chip sensors, error propagation is prevented by the immediate discarding of the abnormal droplet. The biochip implements error-recovery operations in subsequent operations.

The closed-loop integration in cyberphysical microfluidics can also be used to control the completion time of bioassays. For example, when measuring the glucose in blood, serum samples need to be well-mixed with an enzymatic reagent [19]. During the mixing procedure, the status of the droplet is monitored by an image sensor. The extent to which mixing has been completed can be quantified by analyzing and comparing images of the droplet. Therefore, the control software will force the biochip to continue mixing until the feedback information shows that the droplets are sufficiently well-mixed. Hence without knowing the precise mixer execution time in advance, the mixing operation can be precisely controlled, and the on-chip measurement results for glucose concentration can be as precise as the results derived by a traditional bench-top analyzer used by a laboratory technician [19].

2.3   Reliability-Driven Error-Recovery

2.3.1   Error Recovery Strategies

In this subsection, we formulate the principles underlying error-recovery. For the given bioassay protocol, we use the sensing system on-chip to evaluate the quality of output droplets of each dispensing, mixing, dilution and splitting operation. It is important to note that, the time cost for adding detection operations is negligible, as the response time of on-chip sensors are in the scale of picoseconds or nanoseconds [13].

One a microfluidic biochip, there are two categories of fluid-handling operations: reversible and nonreversible operations. Reversible operations include dispensing and splitting operations; nonreversible operations include mixing and dilution operations. For errors that occur at reversible operations, their recovery processes are relative simple. In a splitting operation, if two droplets with unbalanced volumes are generated, then the biochip will first merge the two abnormal droplets to a larger one and then split the larger droplet again. For errors that occur at a dispensing operation, the chip can send the abnormal droplet back to the corresponding reservoir and dispense another droplet. Thus for errors that occur at reversible operations, the time cost for recovery is low and no additional droplets need to be consumed.

The error-recovery process for nonreversible operations is more complicated. In order to re-execute the corresponding nonreversible operations to correct the error, we also need input droplets from operations whose outputs feed the inputs of the failed operation. Thus we may need to re-execute all the predecessors of the erroneous operation. For instance, if an error occurs at operation 7 in Fig. 2.6a, operations 1, 2, 3, 4, 5, and 6 may need to be re-executed. Thus the time cost for executing error-recovery operations can be extremely high. The following strategies are taken in our approach to reduce the incidence of the worst case:

- For a splitting operation, if only one of its output droplets is used as the input for the immediate successors, the other (redundant) droplet will be stored as a backup for possible error-recovery at a subsequent stage. For example, operation 7 in Fig. 2.6a is a splitting operation and it generates two output droplets. Only one of these two droplets is used as the input of operation 9. (Note here each circle in the sequencing graph stands for a fluid-handling operation. The unused droplets are not shown in the sequencing graph.) If an error occurs at operation 9, the redundant droplet will be used as an input for re-execution.

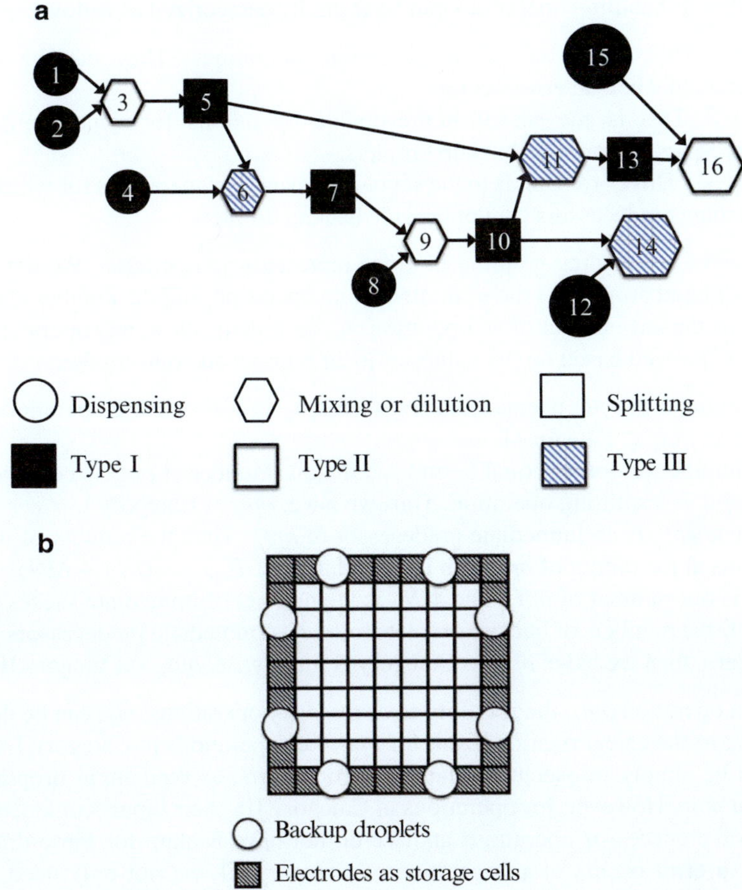

Fig. 2.6 (**a**) An example of a sequencing graph corresponding to a bioassay protocol; (**b**) the layout of a biochip with reserved area for error-recovery

- All dispensing operations are scheduled for execution as early as possible and their output droplets are stored on the biochip. We also dispense some droplets as backup for possible error-recovery operations. After the bioassay is completed, those unused backup droplets will be sent back to their corresponding reservoirs.

Thus, when an error occurs at a nonreversible operation, the control software first checks whether the inputs of this operation can be provided by backup droplets stored on the biochip. If the answer is yes, then the time cost for this operation can be shortened. Otherwise, more operations will be executed during error-recovery. Based on the above discussion, the operations in the bioassay can be divided into three categories according to the number of operations and droplet consumptions in their error-recovery processes, as shown in Fig. 2.6a.

The fluidic-handling operations can be formally categorized as follows:

Category I: This is the set of all reversible operations. They can be simply re-executed when an error occurs.

Category II: This is the set of nonreversible operations for which immediate predecessors can provide backup droplets.

Category III: This corresponds to the set of nonreversible operations for which their immediate predecessors cannot provide backup droplets.

In a given sequencing graph, each node represents an operation. We define the number of input droplets as the in-number of an operation, and the number of output droplets as the out-number of an operation. As described below, any operation opt_k can be categorized based on the values of its in-number and out-number:

- If in-number of opt_k is equal to zero, then opt_k is a dispensing operation. Thus we have: $opt_k \in$ Category I.
- If in-number of opt_k is equal to one and the out-number of opt_k is equal to two, then opt_k is a splitting operation. Thus we have: $opt_k \in$ Category I.
- Suppose opt_j is an immediate predecessor of opt_k. Then the number of backup droplets at the output of opt_j can be calculated as: $B_{opt_j} = ON_j - MN_j$, where ON_j is out-number of opt_j, and MN_j is the number of immediate successors of opt_j. If the numbers of backup droplets for opt_k's immediate predecessors are all non-zero, then we have: $opt_k \in$ Category II; otherwise, $opt_k \in$ Category III.

For an operation opt_i, the set of its error-recovery operations, $\mathcal{R}_i$, can be derived according to the categorization result for opt_i. For operations in Category I and II, they can be simply re-executed when an error occurs, as their input droplets are stored on chip. However, for operations in Category III, their inputs come from the outputs of predecessor operations and we do not have backup for these droplets. Thus if an error occurs in an operation of Category III, we not only need to re-execute the operation itself but also need to backtrace to its predecessors. Assume that the error operations is opt_e and its immediate predecessors are operation opt_{p_1} and opt_{p_2}. If these immediate predecessors are operations in Category I or Category II, we can first re-execute opt_{p_1}, opt_{p_2} and then opt_e for error-recovery, thus $\mathcal{R}_i = \{opt_{p_1}, opt_{p_2}, opt_e\}$. If the immediate predecessors opt_{p_1} and opt_{p_2} are neither in Category I nor Category II, we have to continue to enlarge $\mathcal{R}_i$ by adding the immediate predecessors of opt_{p_1} and opt_{p_2} into $\mathcal{R}_i$. This backtracing and enlargement procedure needs to be repeated until we reach predecessor operations that can provide backup droplets to feed the inputs of operations in the set of error operations.

The above procedure of backtracing and enlargement of the set $\mathcal{R}_i$ can be described as follows. First, we define the mapping $pred(opt_i)$ to be a mapping from opt_i to the set of immediate predecessors of opt_i in the sequencing graph.

For a set of operations $\mathcal{O} = \{opt_{o_1}, opt_{o_2}, \ldots, opt_{o_k}\}$, we define the operator $\mathcal{P}_r$ as:

1: Classify operations into Category I, Category II and Category III;
2: Initialization of $\mathscr{R}_i$: $\mathscr{R}_i = opt_i$;
3: Initialization of intermediate variable Re: $Re = \mathscr{P}_r(\mathscr{R}_i)$;
4: **while** $(Re - \mathscr{R}_i) \cap \{$Set of operations in Category III$\} \neq \emptyset$ **do**
5:     Update $\mathscr{R}_i$: $\mathscr{R}_i = \mathscr{P}_r(\mathscr{R}_i)$;
6:     Update Re: $Re = \mathscr{P}_r(\mathscr{R}_i)$;
7: **end while**
8: $\mathscr{R}_i = Re$;
9: $\mathscr{R}_i$ is the set of recovery operation for opt_i;

Fig. 2.7 Pseudocode for determining the recovery operation for opt_i

$$\mathscr{P}_r : \mathscr{O} \rightarrow \bigcup_{i=o_1,o_2,\ldots o_k} \{opt_i, opt_j \,|\, opt_j \in pred(opt_i), \forall\, j\}$$

For any operation opt_i, its set of error-recovery operations $\mathscr{R}_i$ can be derived by the procedure presented in Fig. 2.7. According to the above discussion, for any operation opt_i we can derive the set of recovery operations $\mathscr{R}_i$.

Based on the relationship between operations in the initial sequencing graph, we can further add edges between operations in the set $\mathscr{R}_i$, and thus derive the error-recovery graph G_{Re_i} for opt_i. If an error occurs in opt_i, we will re-execute operations in G_{Re_i} for error-recovery.

It is important to note that some electrodes on the biochip are intentionally left unused and reserved for storage of backup droplets. An example is shown in Fig. 2.6b; all electrodes on the boundary of the chip are allocated and reserved as storage cells. Thus backup droplets can be easily transported on the biochip.

2.3.2 Reliability Consideration in Error-Recovery

When an error is detected during the execution of a bioassay, it is inefficient to ensure reliable operations by simply re-executing the operation for which an error occurred. This is because the errors that occur during the execution of a bioassay usually are caused by defects involving electrodes; thus, multiple errors may occur in the same region of the biochip at different times. Two examples are discussed below to illustrate the errors caused by the charge-trapping phenomenon and DNA fouling.

When the electrodes of a digital microfluidic biochip are actuated excessively, physically-trapped charge and residual charge may lead to reliability problems [7,8]. Charge trapping is a phenomenon in which charge is trapped and concentrated in the dielectric insulator of the biochip. The trapped charge can lead to a reduction in the electrowetting force and malfunctions in the execution of the bioassay. An example is shown in Fig. 2.8a.

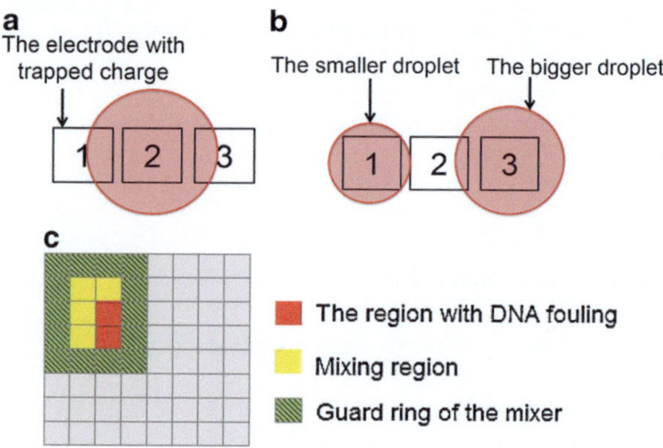

Fig. 2.8 (**a**) An error caused by the phenomenon of charge trapping; (**b**) splitting operation with droplets with unbalanced volumes; (**c**) an error caused by DNA fouling on the surface of a biochip

Suppose Electrode 1 has trapped charge in its dielectric insulator layer, while Electrode 2 and 3 do not suffer from trapped charge. In order to implement a splitting operation, actuation voltages are applied on Electrode 1 and Electrode 3. However, the charge trapped on Electrode 1 will weaken the electrowetting force. The droplet will be split by unequal forces, and the two resulting droplets may have unequal volumes; see Fig. 2.8b. If we simply re-execute the splitting operation and continue to use Electrode 1, additional errors may result. Even worse, the charge-trapping phenomenon may eventually cause permanent dielectric degradation of the electrode [7, 8]. Thus, in order to ensure the reliability of the biochip, the electrode at which the error occurred must no longer be used to implement fluid-handling operations once an error is detected.

The droplets containing macromolecules (such as DNA) may foul the surface of the electrodes [20]. As a result, droplet concentration can change in undesirable ways. If we continue to use these contaminated electrodes, other droplets may also be contaminated. An example of this is shown in Fig. 2.8c. The region where DNA fouling occurred is used as part of a mixer, and the output droplets of the mixing operation may have abnormal concentrations.

We use a simple strategy to ensure a reliability-driven error-recovery. When an error is detected, we update the execution of the bioassay as follows:

- The operation with error is re-executed.
- The electrodes that may lead to errors will not be used in other operations. Note that the on-chip resources occupied by each operation are recorded by the control software. Thus depending on the error droplet, it is feasible to backtrace to the region where error occurs. We consider all electrodes in this region as the suspicious locations for defects. These electrodes will therefore be bypassed.

Table 2.2 Synthesis results for the bioassay shown in Fig. 2.1a

Operation	Start time	Stop time	Resource	Location
Mix 1	6	12	3×2 mixer	(2, 6)
Mix 2	0	6	2×3 mixer	(2, 5)
Mix 3	0	10	2×2 mixer	(6, 2)
Mix 4	12	15	4×4 mixer	(4, 6)
Mix 5	15	18	4×2 mixer	(4, 6)

Table 2.3 Synthesis results for the bioassay shown in Fig. 2.1a

Operation	Start time	Stop time	Resource	Location
Mix 1	6	12	3×2 mixer	(2, 6)
Mix 2	0	6	2×3 mixer	(2, 5)
Mix 3	0	10	2×2 mixer	(6, 2)
Mix 4	12	15	4×4 mixer	(4, 6)
Mix 5	15	18	4×2 mixer	(4, 6)

2.3.3 Comparison Between Two Sensing Schemes

The two sensing schemes introduced in Sect. 2.2.1 have differences in the context of fault diagnosis, error-recovery, and dynamic re-synthesis.

The diagnosis of an electrode with trapped charge can be used to illustrate the difference between these two sensing systems. Suppose a splitting operation with unbalanced droplets occurs, as shown in Fig. 2.8. In the CCD camera-based sensing system, Electrode 1 can easily be identified as the electrode with residual charges because the droplet volume is smaller than normal volume.

On the other hand, for the optical detector-based sensing system, the outputs of splitter, which consists of Electrodes 1, 2 and 3 shown in Fig. 2.8b will not be used any more. In contrast to the diagnosis result of CCD camera-based sensing system, Electrode 2 and 3 can no longer be used, leading to wastage of on-chip resources.

Next, we use the bioassay shown in Fig. 2.1 to further illustrate the differences of these two sensing schemes.

Suppose the droplets for dispensing operations 1 to 6 in Fig. 2.1a are generated from different dispensing ports. For for all the mixing operations of the bioassay shown in Fig. 2.1a, their synthesis results are shown in Table 2.3. The module placement result corresponding to synthesis result in Table 2.3 can be found in Fig. 2.9.

It is important to note that, in Table 2.3, "resource" refers to part of the electrode array occupied by the mixing operation. The location of a mixer is expressed in terms of the location of the electrode at the upper left corner of the mixer. For example, the upper left corner of the mixing module M_1 is in the sixth row and second column; it includes an electrode array with 2 rows and 3 columns. Thus the mixer is described as a 3×2 mixer at the location (2, 6).

Suppose that in the operation Mix 3 shown in Table 2.3, the DNA-fouling phenomenon occurs after the operation has been underway for 3 s. For the optical

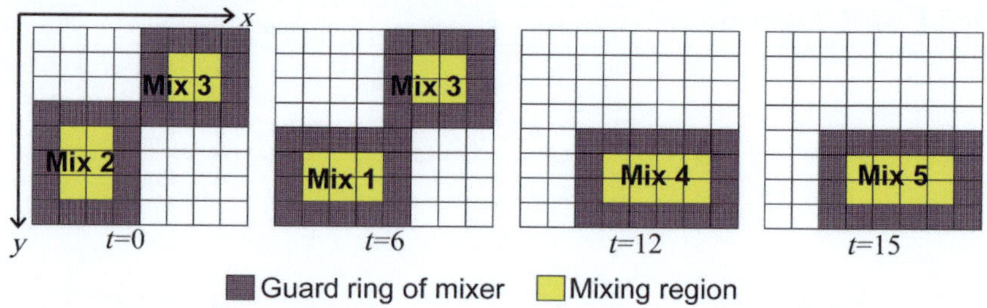

Fig. 2.9 Module placement for the bioassay shown in Fig. 2.1a

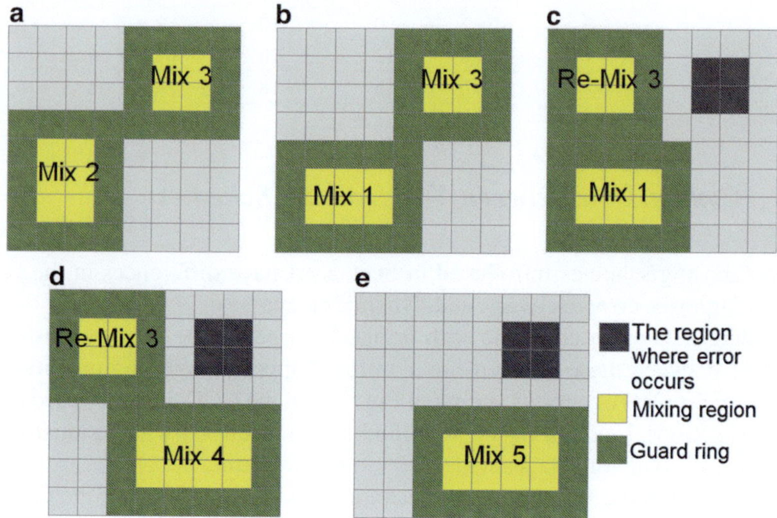

Fig. 2.10 Synthesis results for the bioassay when we use the optical detector-based sensing system. (**a**) $t = 0$: Mix 2 and Mix 3 begin; $t = 2$: DNA fouling occurs at Mix 3 while it will continue to be executed; $t = 6$: Mix 2 is completed; (**b**) $t = 6$: Mix 1 begins. Mix 3 is still being executed (even though DNA fouling has already occurred); (**c**) $t = 10$: Mix 1 is being executed while the output of Mix 3 is sent to optical detector. The error is detected and corresponding electrodes are discarded. $t = 12$: Mix 1 is completed; (**d**) $t = 12$: Mix 4 begins and Re-Mix 3 is being executed; $t = 15$: Mix 4 is completed; $t = 20$: Re-Mix 3 is completed; (**e**) $t = 20$: Mix 5 begins; $t = 23$: Mix 5 is completed. The whole bioassay is completed at time 23

detector-based sensing system, the output of Mix 3 is checked only after Mix 3 has been completed. Thus, the error-recovery process is triggered at time instant $t = 10$. For the CCD camera-based system, the error-recovery process will be triggered immediately after DNA fouling occurs at time instant $t = 3$. The synthesis results for these two cases are shown in Figs. 2.10 and 2.11, respectively. We find that in a detector-based sensing system, recovery can only be triggered at the end of

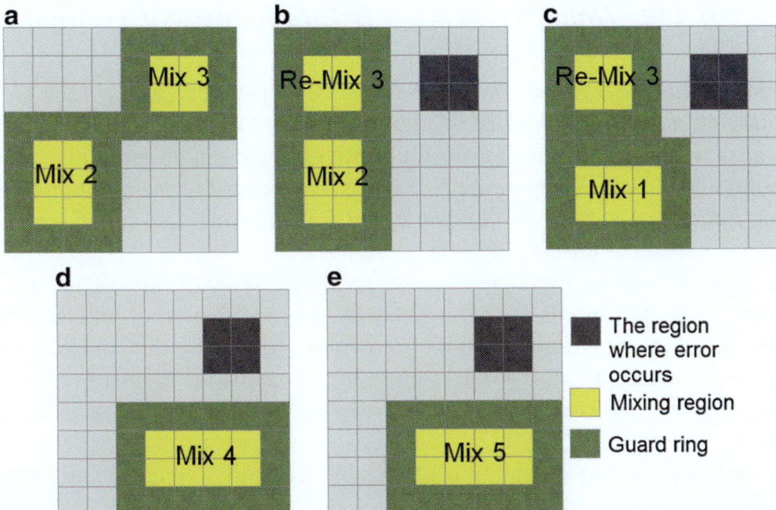

Fig. 2.11 Synthesis results for the bioassay when we use the CCD camera-based sensing system. (**a**) $t = 0$: Mix 2 and Mix 3 begin; $t = 2$: DNA fouling occurs, Mix 3 stops and will be re-executed; corresponding electrodes are discarded; (**b**) $t = 2$: recovery operation Re-Mix 3 begins; $t = 6$: Mix 2 is completed; (**c**) $t = 6$: Mix 1 begins and Re-Mix 3 is still being executed; $t = 12$: both Mix 1 and Re-Mix 3 are completed; (**d**) $t = 12$: Mix 4 begins; $t = 15$: Mix 4 is completed; (**e**) $t = 15$: Mix 5 begins; $t = 18$: Mix 5 is completed. The whole bioassay is completed at time 18

the erroneous operation. While in the CCD camera-based sensing system, recovery can be triggered immediately after an error occurs. On the other hand, in the CCD camera-based sensing system, recovery can only be triggered immediately after an error occurs.

It is important to note that, light from the camera may influence some biochemical substances, e.g., fluorescent markers in the droplet [21]. Thus in order to monitor experiments that include photosensitive samples/reagents, we need to choose the detector-based sensing scheme.

2.4 Error Recovery and Dynamic Re-synthesis

With the availability of hardware that can send feedback to the control software, it is now necessary to design a physical-aware software that can analyze sensor data and dynamically adjust the synthesis result. Adaptations include updates for the schedule of fluid-handling operations, resource binding, module placement, and droplet routing pathways.

The re-synthesis procedure includes two phases: the first phase is off-line data preparation before the execution of bioassay and the second phase is on-line monitoring for the fluid-handling operations as well as dynamic re-synthesis of the bioassay. Details are presented below.

2.4.1   Off-Line Data Preparation Before Bioassay Execution

The first step in data preparation is converting the sequencing graph of the bioassay to a directed acyclic graph (DAG) and storing the DAG in memory for use by the control software. In this DAG, the vertices represent microfluidic handling operations and the edges represent precedence relations between operations. By performing depth-first search on the graph, the predecessors and successors of any operation can be determined [22].

The second step in data preparation is assigning error thresholds for each operation. These thresholds are determined by the requirement of precision for the bioassay and they are stored as a table in memory for use by the control software. During bioassay execution, if a detection result is outside the range of pre-assigned threshold values, we conclude that an error has occurred at the corresponding operation.

The last step in data preparation is deriving the initial synthesis result for the bioassay. In this procedure, we map the sequencing graph of the bioassay and on-chip resources to the scheduling, resource binding, module placement, and droplet routing results for each operation.

For a sequencing graph consisting of n operations, the synthesis result can be written as the following set:

$$\mathscr{S} = \{M^*_{opt_1}, M^*_{opt_2}, \ldots M^*_{opt_n}\}$$

where $M^*_{opt_i}$, $1 \leq i \leq n$, is the synthesis output for the ith operation opt_i. The element $M^*_{opt_i}$ can be viewed as an ordered 6-tuple:

$$M^*_{opt_i} = \; < ts(opt_i), te(opt_i), x(opt_i), y(opt_i), col(opt_i), row(opt_i) >$$

where ts and te are the start time and end time of the operation, respectively; x and y are the x-coordinate and y-coordinate for the module that implements the operation; col (row) is the quantity of columns (rows) occupied by the operation in the array.

For an arbitrary operation opt_i, the order of elements in the tuple $M^*_{opt_i}$ is defined. Thus we can use $M^*_{opt_i}(\tilde{j})$ to represent the $\tilde{j}$th element in the tuple $M^*_{opt_i}$. For example, the start time of ith operation is written as $M^*_{opt_i}(1)$, the x-coordinate of ith operation is written as $M^*_{opt_i}(3)$, and the number of columns occupied by opt_i is written as $M^*_{opt_i}(5)$.

For simplicity, we use $\mathscr{P}$ to represent the set of all operations in the bioassay; and we use $\mathscr{C}$ to refer to the set of constraints that $\mathscr{S}$ must satisfy, which include:

1. For any pair of operations opt_w and opt_v, if the two open intervals $(M^*_{opt_w}(1), M^*_{opt_w}(2))$ and $(M^*_{opt_v}(1), M^*_{opt_v}(2))$ overlap, i.e.,

$$(M^*_{opt_w}(1), M^*_{opt_w}(2)) \cap (M^*_{opt_v}(1), M^*_{opt_v}(2)) \neq \emptyset,$$

which implies that operations opt_w and opt_v are implemented concurrently. It is important to note that multiple operations cannot share on-chip resources (including dispensing ports and electrodes) at the same time. Thus opt_w and opt_v must satisfy following constraint:

$$(M^*_{opt_w}(3), M^*_{opt_w}(3) + M^*_{opt_w}(5)) \cap (M^*_{opt_v}(3), M^*_{opt_v}(3) + M^*_{opt_v}(5)) = \emptyset,$$

$$\bigcup$$

$$(M^*_{opt_w}(4), M^*_{opt_w}(4) + M^*_{opt_w}(6)) \cap (M^*_{opt_v}(4), M^*_{opt_v}(4) + M^*_{opt_v}(6)) = \emptyset,$$

i.e., their corresponding modules cannot overlap with each other.

2. For any pair of operations opt_w and opt_v, if opt_w is the predecessor of opt_v, then opt_w must be completed earlier than the start time of opt_v, i.e. $M^*_{opt_w}(1) \geq M^*_{opt_w}(2)$.

The completion time of the bioassay can be written as:

$$C_p = \max_{opt_i \in \mathscr{P}} \{M^*_{opt_i}(2)\}$$

Thus the synthesis of the biochip can be viewed as an optimization problem. The inputs are the set of operations $\mathscr{P}$ and the set of constraints $\mathscr{C}$. The target is:

$$\text{minimize: } \max_{opt_i \in \mathscr{P}} \{M^*_{opt_i}(2)\}$$

Previously published computer-aided design methods for digital microfluidic biochips have several proposed algorithms to solve this optimization problem. For example, the PRSA-based synthesis algorithms can be used to quickly derive optimized synthesis results [23].

After the optimized synthesis results are derived, the off-line data preparation step is completed. The bioassay is next executed according to the initial synthesis result, and the next step is the on-line monitoring of droplets.

2.4.2 On-Line Monitoring of Droplets and Re-synthesis of the Bioassay

During the execution of the bioassay, the control software must implement the following steps.

2.4.2.1 Step 1: Error Identification

The error identification procedures for the optical detector-based sensing system
and CCD camera-based sensing system are different. For the detector-based sensing
system, the outputs of each operation are sent to an on-chip detector. The software
compares the detection result with a pre-assigned error threshold after each optical
detector operation. If the optical detection result fails to meet the requirement of
the experiment, we conclude that an error has occurred. The detection of the error
will trigger the error recovery procedure, i.e., the software will dynamically adjust
the synthesis results to re-execute the operation in which an error occurred. The
software will also bypass all the suspicious electrodes at which the error occurred.
For example, assume that the output droplet of operation opt_o fails to meet the
requirement, then we can conclude the defect may exist in the resources which are
assigned to opt_o. In order to ensure the reliability of the subsequent operations, this
region will be bypassed.

For the CCD camera-based sensing system, the software can carry out a real-time
monitoring of all droplets on the biochip. The colors and diameters of the droplets
are detected simultaneously by the CCD camera and evaluated for comparisons.
Thus, error recovery is triggered as soon as an error occurs.

2.4.2.2 Step 2: Update of Sequencing Graph

When an error occurs, the control software determines the required recovery
operations. As mentioned above, when an error occurs during the implementation of
operations, the software will adjust the sequencing of the bioassay according to the
category of operation. If the error occurs during a reversible operation, the recovery
process is simple. If the error occurs during a non-reversible operation, the control
software must search the preceding operations until it finds operations that can provide backup droplets to feed the inputs of the recovery subroutine. The pseudocode
for the adjustment of the sequencing graph is shown in Fig. 2.12. The definition
of error-recovery graph G_{Re_i} for operation opt_i can be found in Sect. 2.3.1. For
errors that occur in the operations of Categories I, II, and III, the updated sequencing
graphs are shown in Fig. 2.13a–c, respectively. The categorization of operations can
be found in Sect. 2.3.1.

1: Derive the graph $G_{original}$ by deleting edges between the erroneous operation and its immediate predecessors in original sequencing graph;
2: Derive the error-recovery graph G_{Re_i} for opt_i;
3: Copy G_{Re_i} and label the nodes with different names;
4: Derive the union graph for G_{Re_i} and $G_{original}$;

Fig. 2.12 Pseudocode for adjustment of the sequencing graph

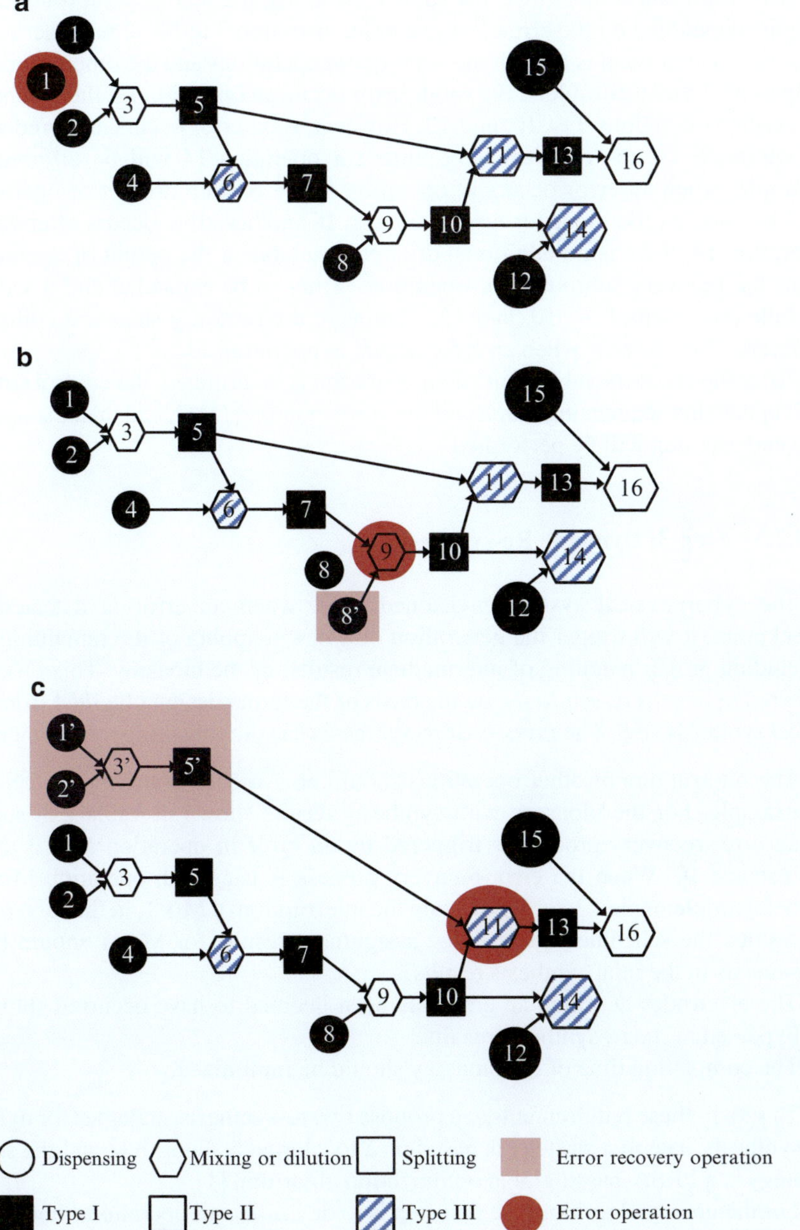

Fig. 2.13 Update of the sequencing graph corresponding to error operations of Categories (**a**) I; (**b**) II; and (**c**) III

It is important to note that, for some operations, the recovery subroutines may change depending on the error. For example, operation 7 in Fig. 2.6a generates two droplets; one of them is used in the subsequent operations and the other is stored on chip as the "backup droplet". If a single error occurs at operation 14, the biochip will re-execute operations $8 \sim 10$, and 12. However, if an error occurs at a predecessor of operation 14, the recovery subroutine for operation 14 will be different. For example, when an error occurs at operation 9, the backup droplet of operation 7 will be used as the input for error-recovery. If another error occurs afterwards at operation 14, there is no additional droplets available at the output of operation 7. Thus the recovery subroutine of operation 14 has to be expanded and it will now include operations $1 \sim 10$, and 12. Therefore the recovery steps are completely different from the case when an error occurs at operation 14.

After the recovery subroutine of an operation is determined, the control software will update the sequencing graph and the corresponding DAG, and then the dynamic re-synthesis step will be performed.

2.4.2.3   Step 3: Dynamic Re-synthesis

In the cyberphysical system envisioned here, when an error is detected at a checkpoint, it will trigger the generation of a new mapping of the remaining steps (including proper handling of intermediate results) of the bioassay. This process is referred to here as *re-synthesis*, on the basis of the initial design obtained from the a priori synthesis step. The process of re-synthesis has the following requirements:

- The interruption of other operations should be avoided. Consider the following example. For the bioassay with synthesis results shown in Table 2.3, suppose an error-recovery process is triggered by an error in operation Mix 3 at time instance 10. When the error-recovery process is triggered, operation Mix 1 is being implemented. In order to avoid the interruption of Mix 1, in the re-synthesis results, the schedule and resource assignment results for Mix 1 should be the same as in the inial synthesis results.
- The electrodes at which an error has been deemed to have occurred should be bypassed in the re-synthesis results.
- The completion time of the bioassay should be minimized.

To satisfy these requirements, we propose two re-synthesis strategies for dynamic re-synthesis. The first strategy is based on a local greedy algorithm, and the second strategy is a PRSA-based global optimization algorithm [1].

For the greedy algorithm, the first step is to determine all operations that must be adjusted in the re-synthesis result. These operations include: the operations in the error-recovery graph, the erroneous operation, and the set of subsequent operations that will be implemented on electrodes with defects in the initial synthesis result. Other operations will be executed based on the initial synthesis result.

Dynamic re-synthesis on the microfluidic array can be modeled as the *module placement with obstacles* problem since the synthesis results for part of the operations are fixed. Here the operations that are implemented based on the initial synthesis result are fixed a priori as the "obstacles". The other operations that are necessary for recovery are derived through re-synthesis and they are placed in the remaining available biochip area in a greedy fashion. The detailed steps are described below.

First, based on the topological sort result for operations, the control software places all operations that need to be re-scheduled in a priority queue. These operations include error-recovery operations and all successors of the erroneous operation. Then the software assigns a priority for each operation in the queue. The "deepest" operation in the subroutine (i.e., the operation at the bottom of the list generated by topological sort) is assigned the lowest priority while the "shallowest" (at the top of the list produced by topological sort) operation is assigned the highest priority in the queue.

Next the control software allocates on-chip resources to these operations. The on-chip resource set R changes with time t. The control software will search for available resources at the current time for the operation with the highest priority. For example, if the operation with the highest priority is a mixing operation, then the system will search for an available $m \times n$ electrode sub-array that is not occupied from current time t to $t + \triangle t$. Here $\triangle t$ is the time needed for the operation in the $m \times n$ electrode sub-array. If suitable idle resources are available, resource binding will be successful and the start time of the operation will be deemed to be the current time. Otherwise, the operation has to be delayed until there are available resources. If multiple resources are available at the same time, the control software will randomly choose one and bind it to the corresponding operation. After binding the resource and determining the start/stop time, the operation will be removed from the priority queue.

Note that when multiple errors are detected at the same time, the above steps can also be used to generate re-synthesis results. In this situation, multiple recovery processes are triggered at the same time and the control software generates a priority queue for each recovery process. After these priority queues are merged, the control software assigns a priority for each element based on topological sort. Finally, the control software determines new synthesis results for every operation in the merged priority queue.

For a microfluidic biochip with an $M \times N$ electrode sub-array and P dispensing ports, the computational complexity of looking for available resources (i.e. "the maximum empty rectangle") in this re-synthesis algorithm is $O(MN + P)$. This is because the software will exhaustively search each electrode/dispensing port in the array and check whether it is available. As the number of dispensing ports can be viewed as constant and we are interested in algorithm scalability for large arrays, the worst-case complexity is $O(MN)$. The computational complexity for other parts of the algorithm are all $O(1)$. Hence the overall computational complexity of the re-synthesis algorithm is $O(MN)$. The pseudocode for the re-synthesis procedure is can be found in Fig. 2.14.

1: Localize the fault operation according to feedback at checkpoints;
2: Determine the operations which need to be adjusted and store them into a priority queue Q;
3: Delete all initial synthesis results for operations in Q;
4: **while** $Q \neq \emptyset$ **do**
5:    Search available resource for operation q_0 which has the highest priority in Q;
6:    Remove q_0 from Q;
7: **end while**

Fig. 2.14 Pseudocode for dynamic re-synthesis of the bioassay

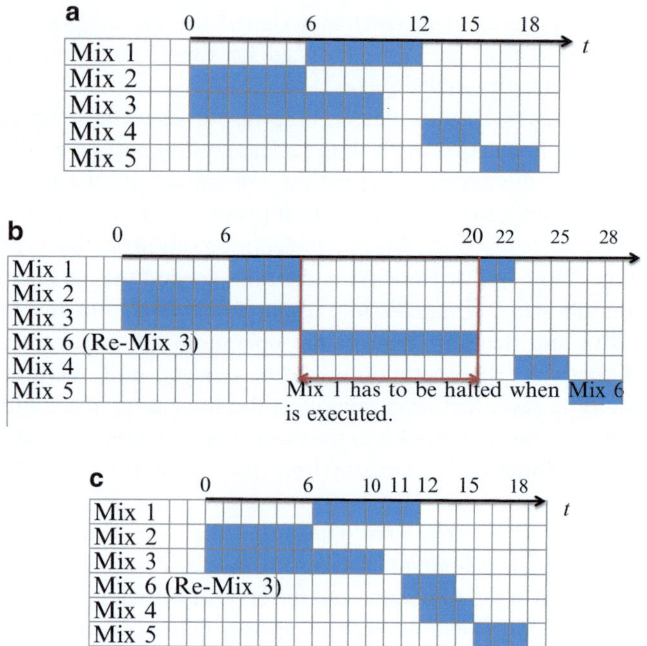

Fig. 2.15 (**a**) Scheduling result when no error occurs; (**b**) scheduling when an error occurs in Mix 3. Mix 1 is halted when error operations are executed; (**c**) scheduling when an error occurs in Mix 3. Here dynamic synthesis strategy is applied at time 10 and error-recovery operations begin at time 11

An example of re-synthesis is shown in Fig. 2.15. Figure 2.15a shows the schedule corresponding to the sequencing graph in Fig. 2.1a. Figure 2.15b, c both show the schedules corresponding to the sequencing graph in Fig. 2.1b. For the sake of clarity, we only show the schedule for mixing operations. Here Fig. 2.15b is the schedule obtained using the error-recovery algorithm of [11].

From Fig. 2.15b, we can see that mixing operation Mix 1 is halted for 10 time slots when error-recovery operations are executed. The completion time of the bioassay shown in Fig. 2.15a increases from 18 time slots to 28 time slots, which can be unacceptable for many applications. The dynamic scheduling result corresponding to Fig. 2.1b is shown in Fig. 2.15c. When the error is detected in the output of Mix 3 at time 10, the ongoing operation Mix 1 is executed based on the initial synthesis result. We assume that the computing time of generating the new synthesis result is 1 time slot. In practice, the computation time is at least an order of magnitude less than the operation time of fluid-handing operations. Then at time 11, the control software will generate re-synthesis result based on the updated Fig. 2.1b. As shown in Fig. 2.15c, Mix 1 is completed at time 12 without being interrupted. The experiment is finished at time 18. Thus the bioassay is executed "seamlessly" without any time penalty or interruption of other operations.

The re-synthesis problem can also be solved using the PRSA-based global optimization method from [1]. The inputs and constraints of the re-synthesis problem are different from the initial synthesis problem introduced in Sect. 2.3.1. Suppose the set of operations for the re-synthesis problem is $\mathscr{P}'$ and the set of constraints is $\mathscr{C}'$. We can derive $\mathscr{P}'$ and $\mathscr{C}'$ based on $\mathscr{P}$ and $\mathscr{C}$ introduced in Sect. 2.3.

We first define an operator $\mathscr{T}$ on the set $\mathscr{P}$. $\mathscr{T}$ is a mapping from the set of all operations to the set of operations that have already started at time instant t.

$$\mathscr{T}(t) : \mathscr{P} \rightarrow \mathscr{P}(t) = \{opt_i | M^*_{opt_i}(1) \leq t\}$$

When an error is detected at time instant t in operation opt_i, the set of operations that need to be re-synthesized can be written as $\mathscr{P}' = \mathscr{P} \cup \mathscr{R}_i \cup \tilde{\mathscr{O}} - \mathscr{P}(t)$. Here $\mathscr{R}_i$ is the set of recovery operations corresponding to erroneous operation opt_i, and $\tilde{\mathscr{O}}$ is the set of subsequently operations which will be implemented on electrodes with defect in the initial synthesis result. The method for determining the operations in $\mathscr{R}_i$ is introduced in Sect. 2.4.1. Then based on the module placement information included in initial synthesis result, and locations of electrodes with defects, operations in $\tilde{\mathscr{O}}$ can be determined.

We write the new synthesis results for $opt_i \in \mathscr{P}'$ as M'_{opt_i}. In addition to the set of constraints $\mathscr{C}$, the re-synthesis result must satisfy the constraint that: the region where an error has been deemed to have occurred, cannot be used any more.

The optimization problem for re-synthesis process can be written as:

$$\text{minimize: } \underset{opt_i \in \mathscr{P}'}{\text{Max}} \{M'_{opt_i}(2)\}$$

The above optimization problem can be solved by using the PRSA-based synthesis procedure introduced in [1]. Using this method, we can derive globally-optimized synthesis results with short assay completion time, while the CPU time is in the order of 20 min for a typical bioassay [11]. Thus, this method is not suitable for on-line computation of re-synthesis results.

2.5 Simulation Results

In this section, we evaluate the re-synthesis approach for error-recovery on representative bioassays that are especially prone to fluidic errors. The completion times for the two sensing schemes are compared; the re-synthesis results derived by the greedy algorithm and the PRSA-based global optimization algorithm are also presented.

2.5.1 Preparation of Plasmid DNA

First, we simulate the bioassay that is called "sample preparation of plasmid DNA by alkalinelysis with SDS" [17, 24]. During sample preparation, a mixture of three reagents is required. The three reagents are:

- R_1: Alkaline lysis Solution I [50 mM Glucose,25 mM Tris–HCl (pH 8.0), 10 mM EDTA (pH 8.0)].
- R_2: Alkaline lysis Solution II [0.2 N NaOH, 1 % SDS (w/v)].
- R_3: Alkaline lysis Solution III (5 M sodium acetate, glacial acetic acid).

The required concentration of the mixture is 0.22 % of R_1, 0.44 % of R_2, and 0.34 % of R_3, which can be approximated as $\frac{28}{128}$ of R_1, $\frac{56}{128}$ of R_2, and $\frac{44}{128}$ of R_3. Figure 2.16 shows the sequencing graph to obtain the required concentration by mixing R_1, R_2, and R_3. This bioassay is mapped to a 10×10 electrode array and all the electrodes at the boundary of the array are used as storage cells.

When errors are detected, the error-recovery capability of the cyberphysical microfluidic system can be evaluated on the basis of the bioassay completion time. The errors are randomly injected into the chip during the execution of the bioassay and compare the completion time of the two sensing schemes. The results are shown in Fig. 2.17. Here the completion time is derived from the greedy algorithm introduced in Sect. 2.4. The results are the average of the values derived from repeating the experiments ten times. For this case (no error-recovery), the final

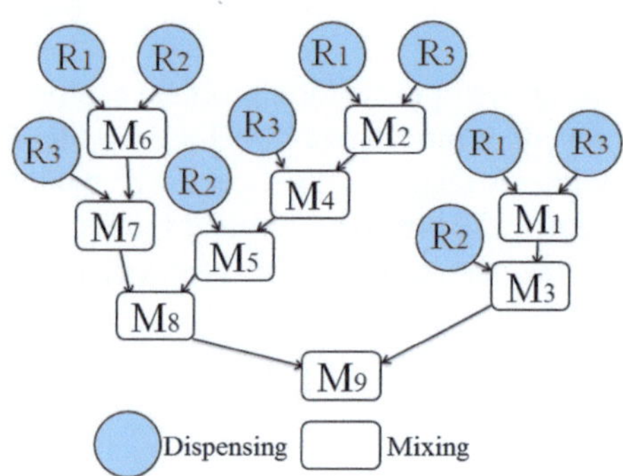

Fig. 2.16 Sequencing graph for sample preparation of plasmid DNA

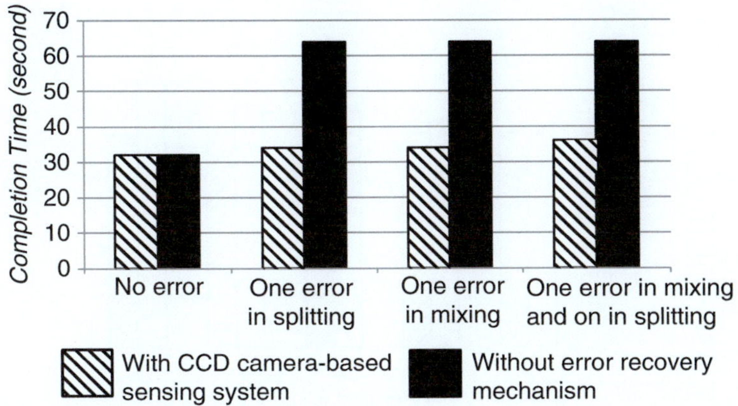

Fig. 2.17 Completion time for biochip with CCD camera-based sensing system, and the biochip without error-recovery mechanism when errors are injected in the sample preparation of plasmid DNA

outcome of the entire experiment will be incorrect if an error occurs during the bioassay. As a result, the biochip has to be discarded, and the experiment must be repeated on a new biochip in order to correct the error. If we assume that re-execution of the experiment will be successful, the bioassay completion time will be twice as the completion time in the fault-free case. Based on these results, we note that error-recovery can reduce the bioassay completion time, and the consumption of biochemical reagents/samples can be reduced.

In reliability-driven error-recovery, the electrodes where an error is deemed to occur, will not be used in other operations. On the contrary, for the reliability-oblivious error-recovery process in [17], when an error occurs during execution, the region where error occurs will continue to be used in subsequent operations. As discussed in Sect. 2.3.2, these electrodes with defects may further lead to more errors.

To compare the completion times derived from reliability-oblivious and reliability-driven error-recovery procedures, the following simulation is set up. In the reliability-oblivious error-recovery, we randomly select one operation opt_{fe} as the first instance of error in the execution of bioassay. The electrodes that are used to perform opt_{fe} are referred to "electrodes with defects". When another operation is implemented again on these electrodes with defects, we assume that there exists a probability P_{fail} that this operation will also fail. For a fixed value of P_{fail}, we simulate reliability-oblivious error-recovery 15 times, and determine average completion time.

Figure 2.18 compares the completion time of reliability-driven error-recovery and average completion time of reliability-oblivious error-recovery for different values of P_{fail}. Here the randomly selected opt_{ef} is a mixing operation implemented on a 4×1 electrode array. As expected, Fig. 2.18 shows that the reliability-driven error-recovery leads to shorter assay completion time in the presence of defects.

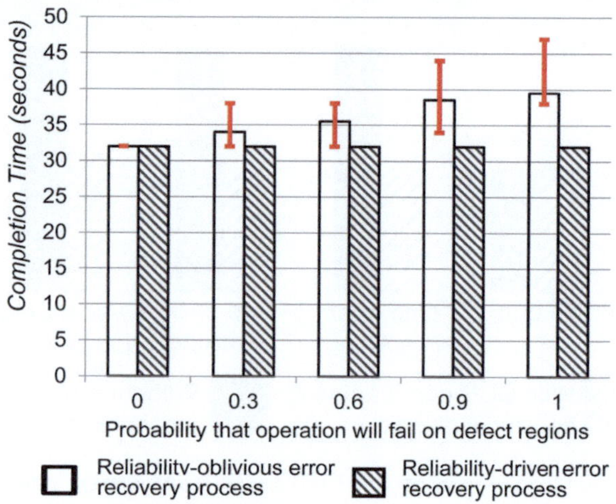

Fig. 2.18 Comparison for the completion time between reliability-driven and reliability-oblivious error-recovery [17] when a 1×4 sub-array is defective in the sample preparation of plasmid DNA. The *error bars* show the maximum and minimum completion time for reliability-oblivious error-recovery in simulation

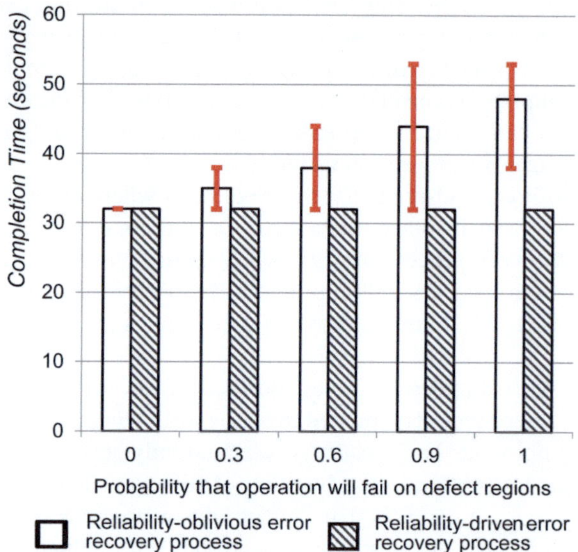

Fig. 2.19 Comparison between the completion time of reliability-driven and reliability-oblivious error-recovery when a 2×4 sub-array is defective in the sample preparation of plasmid DNA. The *error bars* show the maximum and minimum completion time for reliability-oblivious error-recovery in simulation

Next we randomly select another operation as opt_{fe} and run the simulation again. The electrodes that implement opt_{fe} now constitute of a 4×2 electrode array. The simulation results are shown in Fig. 2.19. We find that as expected, the average completion time for reliability-oblivious error-recovery is higher when more electrodes are defective. The completion time of the reliability-driven error-recovery does not depend on the type of defect on the chip and keeps the minimum completion time.

2.5.2 *Protein Assays: Interpolating Mixing and Exponential Dilution*

Next we evaluate re-synthesis and error-recovery for two real-life protein assays. These assays lead to the dilution of a protein sample by using two methods, namely interpolating mixing and exponential dilution. Figure 2.20 shows the sequencing graphs of these two protocols [11]. The protocols for these two bioassays are described in [11].

The completion time of biochips with CCD camera-based sensing systems and without error-recovery mechanism are shown in Fig. 2.21 when errors are injected in the sample preparation of interpolating mixing. The bioassay is mapped to a 10×10 electrode array.

Figure 2.22 reports the completion time when multiple errors are inserted into the interpolating mixing bioassay. Note that the completion time defined here only includes the time spent on fluid-handling operations, and excludes the CPU time consumed on resynthesis. From Fig. 2.22, we see that the completion time achieved by the PRSA-based algorithm and the greedy algorithm are almost the same, but the CPU times for these two algorithms are different. The simulation is performed on a 2.6-GHz, Intel i5 processor with 6 GB of memory. Both re-synthesis algorithm are implemented on the basis of the same initial synthesis result. The CPU time needed is around 33 min for computing the re-synthesis results using PRSA, which was ten times higher than the bioassay completion time; while the CPU time is less than 5 s for the greedy algorithm, which is only 2.5 % of the bioassay completion time. The bioassay completion time derived by the greedy algorithm is only slightly higher for the PRSA. Nevertheless, the greedy algorithm is more suitable for on-line re-synthesis due to the low CPU time.

While the PRSA-based approach is less attractive for real-time decision making, it provides a useful calibration point for the greedy algorithm and shows that the latter's effectiveness for timely bioassay completion. Moreover, the PRSA-based method can be served as the basis for future error-recovery methods based on pre-computation and pre-loading of recovery schedules.

For the exponential dilution protocol introduced in [11], we compare the completion time for the reliability-driven and reliability-oblivious error-recovery methods in Fig. 2.23. First we randomly select one operation opt_{fe} as the first instance of error in the execution of bioassay, where opt_{fe} is a dilution operation performed on a 1×4 electrode sub-array. Then for subsequent operations that are performed on this electrode array with defects, we set P_{fail} as the probability that the operation will fail

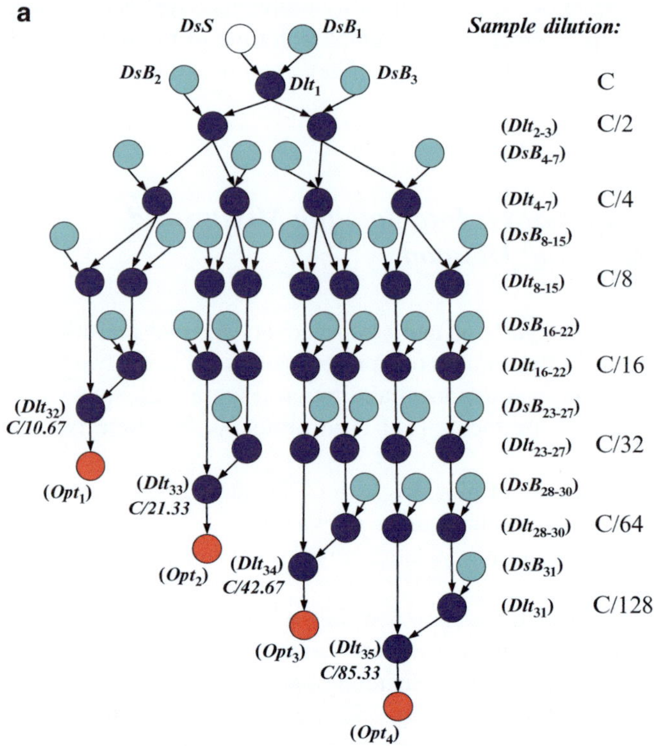

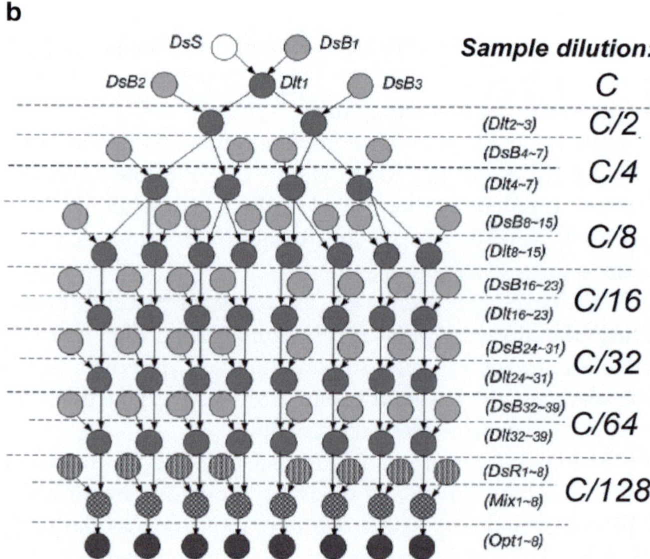

Fig. 2.20 Sequencing graphs for (**a**) interpolating mixing assay; (**b**) exponential dilution of a protein sample [11]

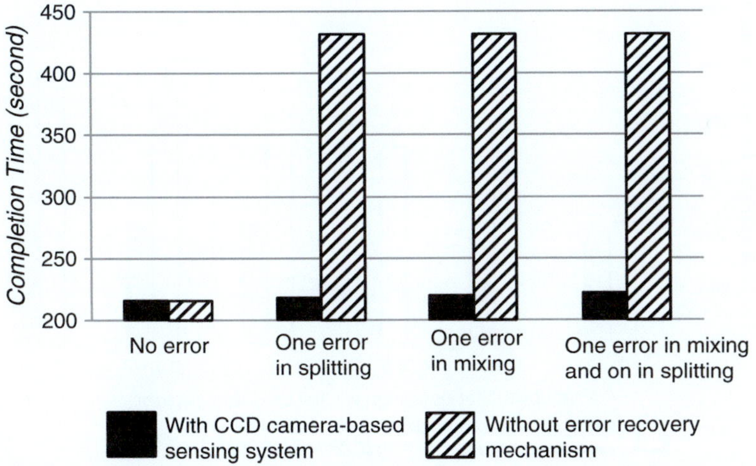

Fig. 2.21 Completion time for biochips with CCD camera-based sensing systems and without error-recovery mechanism when errors are injected in the sample preparation of interpolating mixing

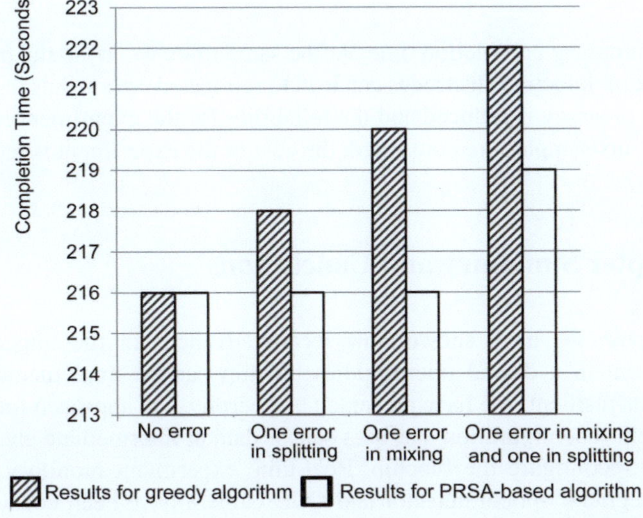

Fig. 2.22 Completion time for the bioassay of interpolating mixing (derived from two re-synthesis algorithms when multiple errors are injected)

again. Then corresponding to each value of P_{fail}, we run the simulations 15 times, and derive the average completion time for reliability-oblivious error-recovery. In contrast, the defective electrodes are bypassed in reliability-driven error-recovery. Thus the completion time of reliability-driven error-recovery is independent of P_{fail}. From the results shown in Fig. 2.23, we find that reliability-driven error-recovery

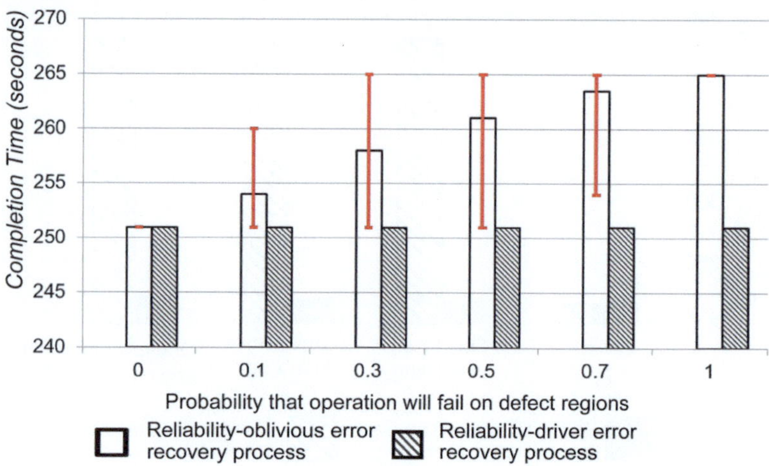

Fig. 2.23 Comparison between the completion time of reliability-driven and reliability-oblivious error-recovery [17] when a 1 × 4 defect array is injected in exponential dilution. The *error bars* show the maximum and minimum completion time for reliability-oblivious error-recovery in the simulation

reduces the bioassay completion time. At the same time, we avoid the problem that any given set of defective electrodes can lead to replicated errors, thus the number of errors in the bioassay is reduced and the reliability for the experiment is improved. As less reagents/samples are consumed, the cost of the experiment is reduced.

2.6   Chapter Summary and Conclusions

In this chapter, we have shown how recent advances in the integration of a sensing system in a digital microfluidics biochip can be implemented to make biochips error-resilient. We have presented a cyberphysical approach for "physical-aware" system reconfiguration that uses sensor data at intermediate checkpoints to dynamically reconfigure the biochip. Real-time experiment monitory techniques based on integrated optical detector and CCD camera have been considered. Two different sensor-driven re-synthesis techniques have been developed to dynamically generate new schedules, module placements, and droplet routing pathways for the bioassay, with minimum impact on the time-to-response. These two methods have been evaluated and compared in terms of bioassay completion time and CPU time needed for re-synthesis. The coordination between the physical-aware control software and the microfluidic biochip allows sensor data to be used as feedback to make decisions about completed operations, to optimize electrode actuation sequences for subsequent operations, and to dynamically reconfigure the biochip. The proposed approach has been evaluated through simulation and its effectiveness demonstrated for three representative protein bioassays.

References

1. K. Chakrabarty and F. Su, *Digital Microfluidic Biochips: Synthesis, Testing, and Reconfiguration Techniques*, Boca Raton, FL: CRC Press, 2006.
2. T.-W. Huang, C.-H. Lin, and T.-Y. Ho, "A contamination aware droplet routing algorithm for the synthesis of digital microfluidic biochips", *IEEE Transactions on Computer-Aided Design of Integrated Circuits and Systems*, vol. 29, no. 11, pp. 1682–1695, 2010.
3. E. Maftei, P. Pop, and J. Madsen, "Routing-based synthesis of digital microfluidic biochips", *Proceedings of the 2010 International conference on Compilers, Architectures and Synthesis for Embedded Systems*, pp. 41–50, 2010.
4. T.-W. Huang and T.-Y. Ho, "A two-stage ILP-based droplet routing algorithm for pin-constrained digital microfluidic biochips", *IEEE Transactions on Computer-Aided Design of Integrated Circuits and Systems*, vol 30, no. 2, pp. 215–228, 2011.
5. M. Iyengar and M. McGuire, "Imprecise and qualitative probability in systems biology", *International Conference on Systems Biology*, 2007.
6. O. Levenspiel, *Chemical Reaction Engineering*, New York: Wiley, 1999.
7. J. Verheijen and M. Prins, "Reversible electrowetting and trapping of charge: model and experiments", *ACS J. Langmuir*, No. 15, pp. 6616–620, 1999.
8. J. Park, S. Lee, and L. Kanga, "Fast and reliable droplet transport on single-plate electrowetting on dielectrics using nonfloating switching method", *Biomicrofluidics*, vol. 4, Issue. 2, pp. 1–8, 2010.
9. E. Welch, Y.-Y. Lin, A. Madison, and R. Fair, "Picoliter DNA sequencing chemistry on an electrowetting-based digital microfluidic platform", *Biotech. J.*, vol. 6, pp. 165–176, 2011.
10. S. Kotchoni, E. Gachomo, E. Betiku, and O. Shonukan, "A home made kit for plasmid DNA mini-preparation", *African J. Biotech.*, vol. 2, pp. 88–90, 2003.
11. Y. Zhao, T. Xu, and K. Chakrabarty, "Integrated control-path design and error recovery in digital microfluidic lab-on-chip", *ACM JETC*, vol. 3, no. 11, 2010.
12. C. Mein, B. Barratt, M. Dunn, T. Siegmund, A. Smith, L. Esposito, S. Nutland, H. Stevens, A. Wilson, M. Phillips, N. Jarvis, S. Law, M. Arruda, and J. Todd, "Evaluation of single nucleotide polymorphism typing with invader on PCR amplicons and its automation", *Genome Res.*, vol. 10, pp. 330–343, 2000.
13. R. Fair, "Digital microfluidics: Is a true lab-on-a-chip possible?", *Microfluidics and Nanofluidics*, vol. 3, pp. 245–281, 2007.
14. W. Bialek and J. Onuchic, "Protein dynamics and reaction rates: mode-specific chemistry in large molecules?", *Proceedings of the National Academy of Sciences of the United States of America*, vol. 85, pp. 5908–5912, 1988.
15. N. Jokerst, L. Luan, S. Palit, M. Royal, S. Dhar, M. Brooke, and T. Tyler II, "Progress in chip-scale photonic sensing", *IEEE Trans. Biomedical Circuits and Sys.*, vol. 3, pp. 202–211, 2009.
16. R. Evans et. al., "Optical detection heterogeneously integrated with a coplanar digital microfluidic lab-on-a-chip platform", *Proc. IEEE Sensors Conf.*, pp. 423–426, Oct. 2007.
17. Y. Luo, K. Chakrabarty, and T.-Y. Ho, "A cyberphysical synthesis approach for error recovery in digital microfluidic biochips", *Proc. DATE*, pp. 1239–1244, 2012.
18. Y. Zhao and K. Chakrabarty, "Digital microfluidic logic gates and their application to built-in self-test of lab-on-chip", *IEEE Transactions on Biomedical Circuits and Systems*, vol. 4, pp. 250–262, 2010.
19. B. Hadwen, G. Broder, D. Morganti, A. Jacobs, C. Brown, J. Hector, Y. Kubota, and H. Morgan, "Programmable large area digital microfluidic array with integrated droplet sensing for bioassays", *Lab on a Chip*, pp. 3305–3313, 2012.
20. M. Jebrail and A. Wheeler, "Let's get digital: digitizing chemical biology with microfluidics", *Current Opinion in Chemical Biology*, vol. 14, pp. 574–581, 2010.
21. U. Resch-Genger et. al., "Quantum dots versus organic dyes as fluorescent labels", *Nature Methods*, pp. 763–775, 2008

22. R. Sedgewick, *Algorithms in C: Graph Algorithms*, Boston, MA: Addison-Wesley, Chapter 23, 2001.
23. S. Kirkpatrick, C. Gelatt and M. Vecchi, "Optimization by simulated annealing", *Science*, vol. 220 (4598), pp. 671–680, May 1983.
24. Y.-L. Hsieh, T.-Y. Ho and K. Chakrabarty, "A reagent-saving mixing algorithm for preparing multiple-target biochemical samples using digital microfluidics", *IEEE Transactions on Computer-Aided Design of Integrated Circuits and Systems*, vol. 31, pp. 1656–1669, 2012.

Chapter 3
Real-Time Error Recovery Using a Compact Dictionary

In this chapter, we present a hardware-assisted error-recovery method that relies on an error dictionary for rapid error recovery. The error-recovery procedure and dynamic re-synthesis of a reaction, which is especially attractive for flash chemistry, can be implemented on a single-board microcontroller in real-time. In order to store the error dictionary in the limited memory available in the low-cost microcontroller, we introduce two compaction techniques. We apply five laboratorial protocols to demonstrate that, compared to software-based methods, the proposed dictionary-based error-recovery method has less impact on response time, and requires simple experimental setup, and only a small amount of memory.

This chapter is organized as follows. The research motivation for the hardware-assisted cyberphysical biochip is introduced in Sect. 3.1. The proposed algorithm for creation of the error dictionary is presented in Sect. 3.2. Section 3.3 introduces the generation of actuation matrices corresponding to synthesis solutions in the error dictionary. Section 3.4 illustrates that actuation matrices of bioassays are sparse. Section 3.5 describes the procedures for compaction of the error dictionary. The implementation of dictionary-based error recovery on a field-programmable gate array (FPGA) is introduced in Sect. 3.6. A fault simulation method with consideration of parameter variation in the fabrication process is discussed in Sect. 3.7. Simulation results are shown in Sect. 3.8, and conclusions are presented in Sect. 3.9.

3.1 Motivation and Related Prior Work

Flash chemistry has emerged as a highlight of recent research in chemical synthesis, where reactions are rapidly carried out and carefully controlled in real-time to produce desired compounds with high selectivity [1]. It can be used as a powerful tool for drug discovery, clinical diagnosis, and novel material synthesis [1–3].

© Springer International Publishing Switzerland 2015

Y. Luo et al., *Hardware/Software Co-Design and Optimization for Cyberphysical Integration in Digital Microfluidic Biochips*, DOI 10.1007/978-3-319-09006-1_3

However, flash chemistry introduces the following challenges: (i) organic synthesis procedures require the precise manipulation of liquids in small volumes [1]; (ii) the underlying fast reactions (with time scale of less than a second) require highly precise time-control in each step of chemical synthesis—a requirement that is beyond the capability of today's benchtop laboratory instruments. These challenges can be potentially tackled using miniaturized microfluidic biochips.

Digital (droplet-based) microfluidics is an emerging technology that enables the integration of fluid-handling operations and reaction-outcome detection on a biochip [4] and it is a potential candidate for the implementation of flash chemistry. Liquid droplets with nanoliter or picoliter volumes in a digital microfluidic biochip can be manipulated on an array of discrete unit cells [5]. Fluid-handling operations, such as dilution of samples and reagents [6], transportation of droplets, and crystallization of protein molecules [7], can be implemented on the biochip by applying appropriate voltages to the electrodes [4, 8]. The sequences of actuation voltages are pre-determined (i.e., before the implementation of the fluid-handling operations), and they are stored in a microcontroller or in computer memory [9]. Under clock control, the microcontroller can transfer pre-loaded actuation data to the biochip, thereby making it feasible for laboratory researchers to automatically execute chemical experiments on the biochip. Precise control of the reaction times for each step in the experiment can also be achieved [4]. Furthermore, operations implemented on the biochip can be dynamically reconfigured by reprogramming the actuation sequences [4]. The sequence of actuation voltages can be derived from bioassay protocols using synthesis methods [9–13].

As discussed in Chap. 2, in order to improve the qualify of product droplets, a cyberphysical system implementation of a biochip has been recently proposed [14]. Despite of its novelty and advantages, the error-recovery method proposed in [14] suffers from the following shortcomings:

1. It requires on-line re-synthesis, which involves software-based dynamic regeneration of electrode actuation sequences. Such a resynthesis step leads to increased bioassay response times when errors occur [14]. When on-line re-synthesis is carried out using software, all fluid-handling operations are interrupted. On the other hand, the reaction time for flash chemistry lies in the range of milliseconds to seconds. For example, in Swern-Moffatt-type oxidation, the reaction time is approximately 10 ms at 20 °C [2]. For such organic synthesis, it is essential to precisely control reaction time; these reactions will fail due to the additional time introduced by re-synthesis [1, 3]. Thus the cyberphysical system in [14] cannot be used for flash chemistry.
2. The error recovery approach in [14] must be implemented by the control software running on a computer, which increases the complexity of the cyberphysical system. The computer-in-loop is not always desirable, e.g., for field deployment and handheld devices.
3. The complex nature of the coupling between the biochip and the control software may introduce reliability problems. The communication between the biochip and the software is implemented using three components connected in series:

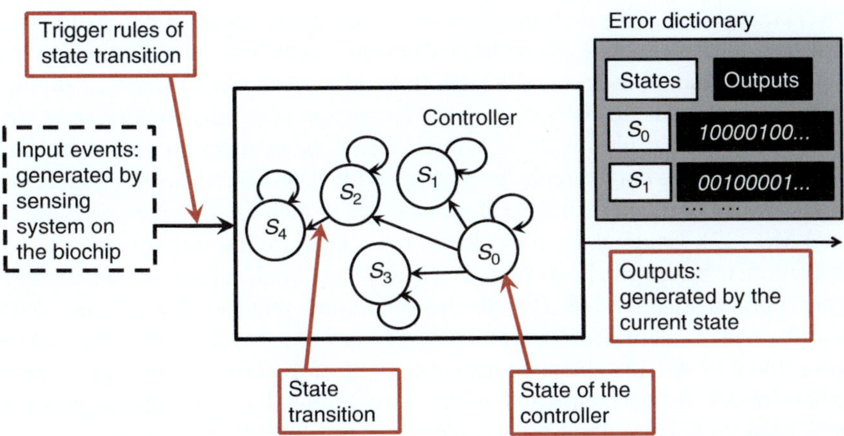

Fig. 3.1 The proposed finite-state machine control system implementation of the cyberphysical system. The FSM runs on the microcontroller or the FPGA, and its current state determines control signals applied on the biochip. A state transition of the FSM is triggered by the detection of error(s)

the peripheral circuit, the microcontroller, and the universal serial bus (USB) port of the computer. Each of these components can be a point-of-failure, resulting in reduced system reliability.

To overcome the above shortcomings, we propose a dictionary-based hardware-assisted error-recovery method. The key idea in this method is to pre-compute and store recovery actuation sequences for all errors of interest that may occur during a bioassay. When an error is detected by on-chip sensors during the execution of a bioassay, the cyberphysical system can simply look up in the dictionary for the recovery solution rather than performing on-line re-synthesis using an in-the-loop computer. This dictionary-based solution therefore reduces response time and enables flash chemistry.

The proposed dictionary-based error recovery approach can be implemented using the finite-state machine (FSM) shown in Fig. 3.1. The control signals for the biochip is determined by the current state of the FSM; the state transition of the FSM is triggered by the analog feedback indicating that an error has occurred. The error dictionary plays the role of a precomputed database that links each possible state to the corresponding outputs. Entries in the dictionary record all possible errors and the re-synthesis results, which include the corresponding error-recovery operations. When an error occurs, the cyberphysical system will utilize the dictionary and load the pre-computed synthesis results.

Since no software execution is needed for on-line error-recovery, the need for a computer and the related interfaces can be eliminated. The dynamic adaptation of the synthesis results can be implemented by a program that executes on the single-board microcontroller or an FPGA, which integrates components such as the microprocessor, programmable I/O interfaces, memory, clock circuitry [15].

However, since the error dictionary needs to store re-synthesis solutions for errors of interest, the data volume can be high. Consider a typical protein dilution bioassay with 103 fluid-handling operations [16]. If we limit ourselves to errors involving at most two operations, the memory required for storage of the error dictionary without compaction can be as high as 28.75 MB. However, the memory on a low-cost FPGA is typically limited. For example, according to the datasheet of several widely-used FPGAs (which cost less than $50), the capacities of their on-chip memories are less than 1 MB [17]. External memory devices can increase the size of total memory, however, their prices can be as high as the low-cost FPGAs [18]. According to the current market price list [18, 19], the use of external memory devices may double the cost of the system. External memory devices require additional circuit-board wiring, and they will also increase the complexity of the system [20]. Therefore, the above solutions are not compatible with the goal of low-cost biochip platforms that can be used for field deployment and point-of-care clinical diagnostics.

In this chapter, we propose a compaction procedure for an error dictionary which includes re-synthesis solutions for all the possible error combinations in bioassay. Then we compact sets of actuation sequences (i.e., actuation matrices) stored in the dictionary in a lossless manner. After these two steps, the size of compacted dictionary in the above example can be reduced to only 0.96 MB.

3.2   Generation of the Error Dictionary

For any given set of errors that can occur in a bioassay, the error dictionary is generated using simulation before the execution of experiments. During simulation, erroneous fluidic operations are considered and the corresponding re-synthesis results are determined. These re-synthesis solutions are stored as entries of the dictionary. Both the simulation and dictionary generation process are performed by a computer in the off-line data-preparation stage.

The error dictionary is stored in controller memory as a decision tree that records all possible consequences of errors, as shown in Fig. 3.2. The dictionary has multiple levels and the entries at the kth level correspond to all possible cases errors that can occur involving k operations. Suppose the total number of operations in the bioassay is N and no errors occur during error recovery. The number of entries in the kth level of the dictionary can be as high as $\binom{N}{k}$. For large N and k, the size of the error dictionary and the CPU time spent on error-dictionary generation can both be prohibitively high. Therefore, k must be limited in practice. Here we consider up to two erroneous operations ($k = 2$) in the bioassay. If more errors occur, the biochip will be discarded and the experiment repeated on a different biochip.

In Sects. 3.2.1–3.2.3, we will introduce the steps for generating the error dictionary level-by-level. First we carry out synthesis for the error-free case, which corresponds to Level 0 in Fig. 3.2; next, based on the error-free solution, we insert errors in the simulation and derive entries in Level 1 and Level 2 of the dictionary.

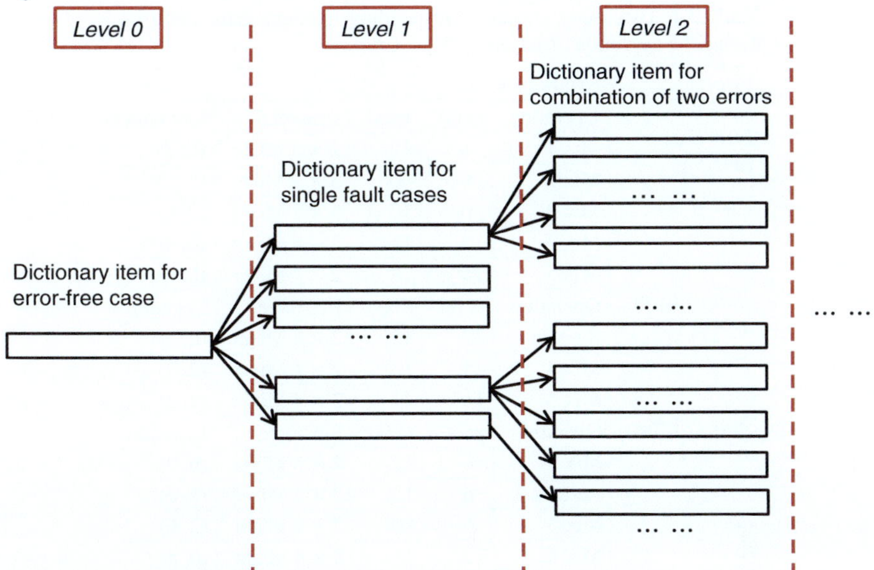

Fig. 3.2 Tree structure of the error dictionary. Entries at the kth level of the dictionary correspond to all possible cases of errors that can occur involving k operations

3.2.1 Dictionary Entry for Error-Free Case

The entry in Level 0 of the error dictionary corresponds to the solution obtained using high-level synthesis. In this process, the behavioral description of the desired bioassay protocol, which is modeled as a sequencing graph, is mapped to a design implementation in terms of a sequence of fluidic operations and the corresponding actuation sequences.

Published computer-aided design methods for digital microfluidic biochips have proposed integer linear programming and heuristic algorithms to solve this optimization problem [10, 16, 21]. For example, the parallel recombinative simulated annealing (PRSA)-based synthesis algorithms can be adapted to derive optimized results [16].

Suppose all the droplets used in the bioassay of Fig. 2.1a are stored on the biochip before the execution of the bioassay. The first entry in Table 3.1 shows the synthesis results derived by PRSA-based synthesis algorithm for all mixing operations of the bioassay shown in Fig. 2.1a. The start and end time of operations are written as *ts* and *te*, respectively. Figure 2.9 shows the module placement corresponding to the synthesis of the error-free case in Table 3.1.

It is important to note that in Table 3.1, "resource" refers to the part of the electrode array occupied by the mixing operation. The location of a mixer is expressed in terms of the location of the electrode at the upper left corner of

Table 3.1 Dictionary entries corresponding to single error occurred during the execution of bioassay

Errors	Synthesis result				
State 0: Error-free case	Operation	ts (s)	te (s)	Resource	Location
	Mix 1	6	12	2 × 3 Mixer	(6, 2)
	Mix 2	0	6	3 × 2 Mixer	(5, 2)
	Mix 3	0	10	2 × 2 Mixer	(2, 6)
	Mix 4	12	15	2 × 4 Mixer	(6, 4)
	Mix 5	15	18	2 × 4 Mixer	(6, 4)
State 1: Error in Mix 1	Operation	ts (s)	te (s)	Resource	Location
	Re-Mix 1	12	15	2 × 4 Mixer	(2, 2)
	Mix 4	15	18	2 × 4 Mixer	(2, 2)
	Mix 5	18	21	2 × 4 Mixer	(2, 2)
State 2: Error in Mix 2	Operation	ts (s)	te (s)	Resource	Location
	Mix 1	6	12	2 × 3 Mixer	(6, 2)
	Re-Mix 2	6	12	3 × 2 Mixer	(2, 2)
	Mix 3*	6	10	2 × 2 Mixer	(2, 6)
	Mix 4	12	15	2 × 4 Mixer	(6, 4)
	Mix 5	15	18	2 × 4 Mixer	(6, 4)
State 3: Error in Mix 3	Operation	ts (s)	te (s)	Resource	Location
	Mix 1*	10	12	2 × 3 Mixer	(6, 2)
	Re-Mix 3	10	20	2 × 2 Mixer	(2, 2)
	Mix 4	12	15	2 × 4 Mixer	(6, 4)
	Mix 5	20	23	2 × 4 Mixer	(6, 4)
State 4: Error in Mix 4	Operation	ts (s)	te (s)	Resource	Location
	Re-Mix 4	15	18	2 × 4 Mixer	(2, 2)
	Mix 5	18	21	2 × 4 Mixer	(2, 2)
State 5: Error in Mix 5	Operation	ts (s)	te (s)	Resource	Location
	Re-Mix 5	18	21	2 × 4 Mixer	(2, 4)

*These operations can be interrupted as part of error recovery

the mixer. For example, the upper left corner of the mixing module M_1 is in the 6th row and 2nd column; it includes an electrode array with 2 rows and 3 columns. Thus the mixer is described as a 2 × 3 mixer at location (6, 2).

In addition, most of the bioassay benchmarks in the literature require that all the input and output droplets for the mixing/dilution operations are droplets with unit volume [22, 23]. Hence at the end of each mixing/dilution operation, the mixed droplet with twice the unit volume must be split into two unit droplets. For each mixing/dilution module, a set of electrodes will be randomly selected as the "splitter" to split the mixed droplet. The splitter can be placed in an arbitrarily-chosen place inside the mixing/dilution module. Therefore, for two operations, even when their corresponding modules are overlapped with each other (e.g., operations Mix 4 and Mix 5 shown in Fig. 2.9), their corresponding splitters may contain different sets of electrodes.

3.2.2   Dictionary Entries for Single-Operation Errors

By inserting an error in the bioassay and recording the corresponding re-synthesis results, we can get the entries in Level 1 of the error dictionary. The re-synthesis problem can also be solved using the PRSA-based global-optimization method [16] while the CPU time is relatively long [14].

In [14], the algorithm for deriving new sequencing graph with error recovery operations is proposed. Based on the synthesis result in the error free case, the newly added operations for error recovery are "inserted" into the synthesis result of the error free case in a greedy fashion [14]. Therefore, the re-synthesis results for error recovery can be generated within limited CPU time.

By explicitly analyzing all possible situations corresponding to all the potential single errors that may occur for a bioassay, we can derive the entries in Level 1 of the error dictionary. For example, the sequencing graph shown in Fig. 2.1a has five mixing operations, and there are five error candidates if we only consider errors in mixing and assume that only one error occurs during the bioassay. We can derive the error-recovery results for all of the potential errors and store the results during the "off-line data preparation" stage, as shown in Table 3.1.

Table 3.1 is loaded into the memory of the FPGA; it determines the state transition of the FSM running on the FPGA. At time $t = 0$, the FSM comes into State 0. The biochip begins to execute the bioassay based on the initial synthesis result. At time $t_s = 6$, operation Mix 2 is completed, the sensing system checks the output of Mix 2. If an error is deemed to have occurred, the FSM transits to the corresponding state (State 2) and loads the re-synthesis result that is stored in the error dictionary (the third entry in Table 3.1). It is important to note that the synthesis results for all operations in the second entry start from time $t = 6$. Since Mix 3 is being executed when an error occurs at Mix 2, it will continue to be executed according to the initial synthesis result. All operations that may be interrupted by the transition for the state of the controller are marked by "*" in Table 3.1. In this manner, the controller can dynamically adapt and do re-synthesis of the on-chip chemistry by looking up entries in the error dictionary; then the computational complexity of on-line re-synthesis of the bioassay is reduced to $O(1)$.

Note that each entry in the dictionary contains only the re-synthesis result from the time moment that the error is detected to the end of the bioassay. For example, the dictionary entry that corresponds to State 1 shown in Table 3.1 records the re-synthesis result from the 12th second (i.e., the moment in time that the error occurs in operation Mix 1 is detected) to the 21st second (i.e., the time moment that the entire bioassay is completed). The time span of this dictionary entry is 9 s. Based on the above discussions, we note that there are various time spans for the entries in Level 1 of the dictionary.

Table 3.2 Synthesis results corresponding to a pair of errors

Errors	Synthesis result				
State (1, 4): Errors in {Mix 1, Mix 4}	Operation	ts (s)	te (s)	Resource	Location
	Re-Mix 4	15	18	2×4 Mixer	(2, 2)
	Mix 5	18	21	2×4 Mixer	(2, 2)
State (1, 5): Errors in {Mix 1, Mix 5}	Operation	ts (s)	te (s)	Resource	Location
	Re-Mix 5	21	24	2×4 Mixer	(6, 4)

3.2.3 Dictionary Entries for Multiple-Operation Errors

Based on the re-synthesis solutions corresponding to single-error cases, we can
further explicitly consider all the cases with more than one error and generate
the re-synthesis resolutions. Consider the dictionary entries shown in Table 3.1.
Assume that an error has already occurred at Mix 1 and the FSM running on the
FPGA has entered State 1. For the re-synthesis solution in State 1, there are three
operations: Mix 4, Mix 5, and Re-Mix 1; here Re-Mix 1 is the recovery operation
for Mix 1. Since we assume that no errors occur in error recovery operations, only
Mix 4 and Mix 5 are considered as candidates for the next error, and we need to
generate recovery solutions for both of them. Finally we derive the re-synthesis
results corresponding to pairs of errors {Mix 1, Mix 4} and {Mix 1, Mix 5}; see
Table 3.2. These solutions are stored as two entries at the second level of the error
dictionary. The "parent node" of these two entries is the entry that corresponds to
the solution for the single error in Mix 1.

Note that each entry that corresponds to the occurrence of multiple errors
contains only the re-synthesis result from the time moment that the latest error is
detected to the time moment that the entire bioassay is finished. For example, the
dictionary entry that corresponds to State (1, 4) shown in Table 3.2 contains two
error operations, i.e., Mix 1 and Mix 4. According to the synthesis results shown
in Table 3.1, the error that occurs in Mix 1 is detected earlier than the error that
occurs in Mix 4. Therefore, the dictionary entry records the re-synthesis result from
the 15th second (i.e., the time moment that the error occurs in operation Mix 4
is detected) to the 21st second (i.e., the time moment that the entire bioassay is
completed). The time span of this dictionary entry is 6 s.

3.2.4 Consideration of Error-Recovery Cost and Reduction in the Number of Dictionary Entries

For low-cost disposable biochips, the cost of samples and reagents used in the
experiments can be higher than the cost of biochips. When an error occurs on
the biochip, the additional number of droplets needed in error recovery needs to

be calculated; then the decision needs to be made on whether to discard the biochip and run the bioassay on a new chip, or recover from the error and continue the experiment.

While generating entries for the error dictionary, an evaluation process can be performed for determining whether an error is "worth being recovered". Thresholds for the number of droplets consumed can be set before generating the error dictionary. Consider the following example. Assume that when there is no error in the bioassay, the number of droplets consumed is N_f, and the maximum number of additional droplets needed for error recovery (N_{max}) is set as $15\% \cdot N_f$. When we generate the dictionary entry for operation O_E, if the simulation result indicates that the number of additional droplets needed for error recovery is more than N_{max}, then the entry for recovering the error in O_E will not be added into the dictionary. Therefore, only cost-efficient entries are recorded by the dictionary. These dictionary entries are referred to as "effective entries" of the error dictionary.

3.3 Actuation Matrix

In order to execute the bioassay on a digital microfluidic biochip, the information of droplet routes and the schedules of operations must be programmed into the controller of the biochip [11]. The synthesis results of the bioassay are mapped to electrode actuation sequences, in which each element represents the status of the electrode at a specific time moment. For an $M \times N$ array, we can number the electrodes on the array as E_1, E_2, $\ldots$, E_{MN}. If the completion time of the synthesis result is T clock cycles, the actuation sequences for all electrodes can be written in the form of an $(M \times N) \times T$ matrix, referred to as the *actuation matrix* and denoted by $\mathscr{A}$. The status of electrode E_i at time j is represented by the element in the ith column and jth row of $\mathscr{A}$.

For an arbitrarily chosen operation $opt_{\bar{i}}$, suppose it is performed on electrodes E_{e_1}, E_{e_2}, $\ldots E_{e_k}$ at clock cycles $T_{t_1}, T_{t_2}, \ldots T_{t_l}$, respectively. If we write the set of the indices of these electrodes as $I = \{e_1, e_2, \ldots e_k\}$ and the indices of clock cycles as $J = \{t_1, t_2, \ldots t_l\}$, then I is a subset of $\{1, 2, \ldots MN\}$ and J is a subset of $\{1, 2, \ldots T\}$. Thus the actuation sequence for $opt_{\bar{i}}$, (which is referred as $M_{opt_{\bar{i}}}$), can be written as $\mathscr{A}_{I,J}$. It is a $k \times l$ sub-matrix of $\mathscr{A}$ that corresponds to the rows with index in set I and the columns with index in set J. According to the constraints of biochemical synthesis, no two operations can occupy the same electrode at the same time, thus for operations $opt_{\bar{i}}$ and $opt_{\bar{k}}$, their corresponding sub-matrices $M_{opt_{\bar{i}}}$ and $M_{opt_{\bar{k}}}$ do not overlap with each other.

Next, we show how the values of elements in the actuation matrices are determined. We also estimate the percentage of non-zero elements in an actuation matrix. During the implementation of fluid-handing operations, the status of the electrodes can be "activated", "deactivated", or "don't-care". A "don't-care" status is assigned to an electrode when it is not required to be either active or inactive. It is important to note that for various operations, the typical control voltages for

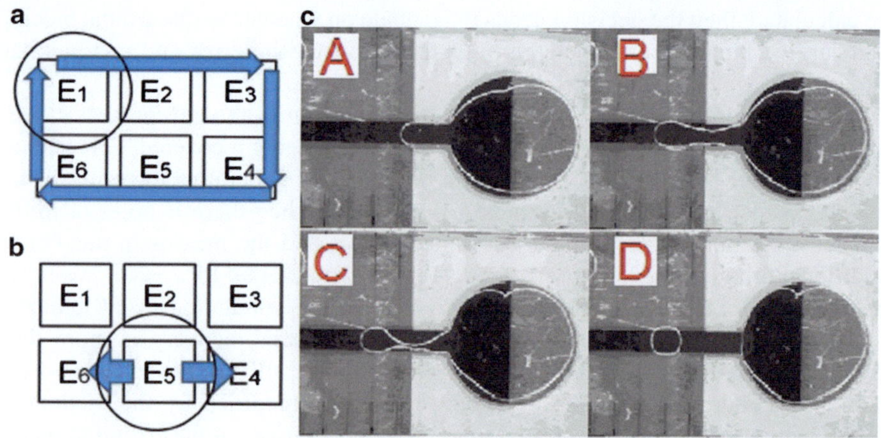

Fig. 3.3 Movement of droplets for (**a**) mixing; (**b**) splitting; (**c**) dispensing [24]. Steps *A–D* show the procedure of pulling a droplet from the reservoir

Table 3.3 Actuation sequences for electrodes in the mixer

Electrodes	1	2	3	4	5	6
Actuation sequences (ordered in time)	v_1	0	X	X	X	0
	0	v_1	0	X	X	X
	X	0	v_1	0	X	X
	X	X	0	v_1	0	X
			...	...		
	X	X	X	v_2	0	v_2

activation statuses are different. For example, in [8], the actuation voltage required to move a droplet is 7.2 V, while the lowest voltage required to dispense a 300 pl droplet from a reservoir is 11.4 V. Thus in the memory of the controller, these non-zero elements of actuation matrices are recorded as the value of the corresponding control voltages. In the following parts, we use "0" and "X" to represent deactivated and "don't-care" status of electrodes, respectively. We also use "v_1" and "v_2" to represent the activated status in terms of different voltages for electrodes in mixing and splitting operations. As shown in Fig. 3.3a, operation Mix 1 is performed on an electrode array with two rows and three columns; at the end of mixing, the product droplet will be split into two smaller droplets, as shown in Fig. 3.3b. During the mixing operation, suppose the droplet is moved counterclockwise along the loop consisting of six electrodes. At each clock cycle, the droplet will be moved from the current electrode to an adjacent electrode. Suppose that at time instant (clock cycle) t_0, the droplet rests on electrode E_1. The actuation sequence for each electrode is calculated and listed in Table 3.3, and the last row in Table 3.3 corresponds to the split operation after the completion of mixing. From Table 3.3, we find that at each time instance during mixing, there is only one electrode in the mixer that must be activated. Voltages applied to other electrodes are either "0" or "X", and all the "X" terms can be replaced by "0".

Next we write the actuation sequences of operation Mix 1 as an actuation matrix M_{Mix_1}. The ith-row vector in M_{Mix_1} indicates the status of E_1 to E_6 at the ith clock cycle of Mix 1. Suppose the completion time for operation Mix 1 is 5 clock cycles and at the 6th clock cycle the mixed droplet will be split. Then M_{Mix_1} has 6×6 elements. It can be written as:

$$
M_{Mix_1} = \begin{bmatrix}
v_1 & 0 & 0 & 0 & 0 & 0 \\
0 & v_1 & 0 & 0 & 0 & 0 \\
0 & 0 & v_1 & 0 & 0 & 0 \\
0 & 0 & 0 & v_1 & 0 & 0 \\
0 & 0 & 0 & 0 & v_1 & 0 \\
0 & 0 & 0 & v_2 & 0 & v_2
\end{bmatrix}.
$$

We see that each row in M_{Mix_1} has one or two non-zero elements, which is a general case in fluid-handling operations.

Figure 3.3c shows the steps involved in dispensing a droplet from an on-chip reservoir. In order to pull a droplet out from the reservoir, the electrodes in the transportation path are activated in sequence, as shown from Steps A to D in Fig. 3.3c [24]. This dispensing procedure is completed in four clock cycles. The droplet is formed by activating the electrodes on the outlet of the dispensing port. In this process shown in Fig. 3.3c, the liquid drop in the reservoir is deformed under the electrical force.

Based on the steps shown in Fig. 3.3c, we derive corresponding actuation matrix M_{Dis} for a dispensing operation. The matrix M_{Dis} has four rows and it includes a total of five non-zero elements.

Since for any two operations $opt_{\tilde{i}}$ and $opt_{\tilde{k}}$, the corresponding sub-matrices $M_{opt_{\tilde{i}}}$ and $M_{opt_{\tilde{k}}}$ are non-overlapping, we can estimate the number of non-zero elements in the actuation matrix $\mathscr{A}$ on the basis of the sub-matrices corresponding to all the operations in the bioassay. If we assume that the working frequency of the biochip is f Hz (f usually varies from 1 to 100 [25]), then an operation that is completed in N seconds includes $N \cdot f$ clock cycles. For the synthesis result shown in the first entry of Table 3.1, the number of non-zero elements corresponding to the sequence matrix of each operation can be calculated accordingly according to their execution time. Apart from the mixing operations, there are six dispensing operations that are not shown in Table 3.1, and each of them corresponds to a sub-matrix with 4 non-zero elements. Thus the total number of non-zero elements in the actuation matrix for the entire bioassay can be estimated by adding the number of non-zero elements for each operation. The resulting number is $24 + 28f$. On the other hand, since the completion time of the bioassay is $4 + 18f$ clock cycles, the size of $\mathscr{A}$ is $(8 \times 8) \times (4 + 18 \cdot f)$. Thus the percentage of non-zero elements (P_{nz}) in matrix $\mathscr{A}$ can be calculated as:

$$P_{nz} = \frac{24 + 28 \cdot f}{(8 \times 8) \times (5 + 18 \cdot f)} \times 100\,\%.$$

As the frequency range is $1 \leq f \leq 100$ [25], we conclude that $2.45\,\% \leq P_{nz} \leq 3.69\,\%$. This implies that $\mathscr{A}$ is a sparse matrix. In Sect. 3.4, we will illustrate that actuation matrices are sparse in general cases.

3.4 Estimation for the Percentage of Non-zero Elements in Actuation Matrices

According to the underlying physical principle of droplet operations, the number of actuated electrodes on the array can be estimated on the basis of the number of droplets that are currently being manipulated on the biochip. The discussion for all the possible fluid-handling operations is as follows.

1. For the droplets that are scheduled to stay at its current position, the electrode under the droplet will be actuated to "anchor" the droplet, while all other electrodes that are in contact with the droplet will be deactivated.
2. For the droplets that are scheduled to be moved from one electrode to another electrode, the target electrodes need to be actuated while all the other electrodes that are in contact with the droplet need to be deactivated.
3. Dilution/mixing operations can be considered as moving two droplets in a module. At each time moment, there are two droplets in a mixing/dilution module and the number of actuated electrodes inside is one or two.
4. Dispensing operations can be considered as "pulling" droplets out of the reservoir. At each time moment, the number of actuated electrodes on the output of the dispensing port is one.
5. For splitting operations, we can consider the $2\times$ droplet before splitting as two $1\times$ droplets. In order to split the $2\times$ droplet, two electrodes need to be actuated. Therefore, the number of actuated electrodes in the splitter is equal to the number of "equivalent" $1\times$ droplets.

Therefore, by calculating the number of maximum number of droplets that can be concurrently manipulated on the biochip, we can find an upper bound on the percentage of non-zero elements in each actuation sub-matrix. Here a sub-matrix represents the set of actuation signals applied on the electrode array at one time moment. For a 10×10 electrode array, each actuation sub-matrix is a 10×10 vector. Next we calculate the maximum degree of parallelism for fluid-handling operations in the exponential dilution of a protein sample. The sequencing graph of the bioassay can be found in Fig. 2.20b [16, 34].

Here we first define the concept of "interdependency operations". If the output of an operation A is the input of operation B, then operations A and B are interdependency operations. It is clear that two interdependency operations cannot be concurrently executed. Therefore, from the sequencing graph, we can find that the maximum number of dilution/mixing operations that can be executed concurrently is 8. When executing these 8 dilution/mixing operations, the biochip may also execute another 8 dispensing operations.

From the above analysis, we can find that the maximum degree of parallelism for fluid-handling operation is: simultaneously implementing 8 mixing operations (and each mixing operation involves two droplets) and 8 dispensing operations (and each mixing operation involves one droplet). Therefore, the maximum number of droplets can be concurrently manipulated on the biochip can be calculated as $2 \times 8 + 8 = 24$. The maximum number of non-zero elements in an actuation sub-matrix is also 24.

As the biochip contains a 10×10 electrode array, the maximum percentage of non-zero elements in actuation sub-matrices is 24 %. Therefore, we can find that 24 % is the upper bound on the percentage of non-zero elements in the error dictionary for the bioassay. It is important to note that, the above analyses can be applied to the situation with any number of errors occurred in the bioassay. This upper bound for the percentage of non-zero elements in an actuation sub-matrix is independent with the number of errors occurred during the execution of bioassay. Therefore, we have shown that, for exponential dilution bioassay, the actuation matrices are sparse.

We can illustrate that actuation matrices are sparse in an alternative way. According to the fluidic constraints of biochip, in order to avoid droplet interferences, each droplet has a guard-ring associated with it. Examples are as follows. Figure 3.4a shows the actuation electrodes for a droplet that stays on an electrode. When the electrode under the droplet is actuated, all the other neighboring electrodes of the droplet need to be deactivated.

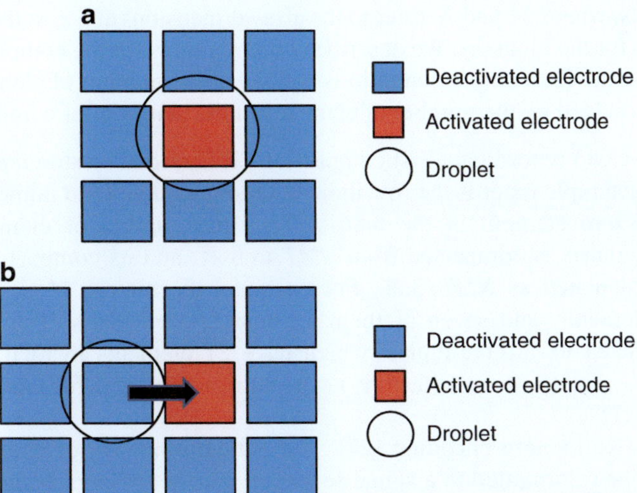

Fig. 3.4 (**a**) The actuation signals applied on electrodes when a droplet stays at an electrode; (**b**) the actuation signals applied on electrodes when a droplet is being moved from one electrode to another electrode

Figure 3.4b shows the actuation signals applied on electrodes when a droplet is being moved. We can find that the destination electrode of the droplet will be actuated, while all the other surrounding electrodes are deactivated.

Similarly, we can analyze the actuation signals that need to be applied on electrodes when a dispensing/mixing/splitting operation is executed. We can find that, usually an actuated electrode is surrounded by a group of deactivated electrodes. When we convert the actuation signals applied on electrodes to a vector or a matrix, we find that non-zero elements (that corresponds to actuated electrodes) are much fewer than zero elements (that corresponds to deactivated electrodes). In this way, we also find that actuation matrices in error dictionaries are sparse.

From the above discussion, we find that the size of the dictionary can be reduced by compacting the actuation matrices. Two data-compaction algorithms for sparse matrices will be introduced in Sect. 3.5.

3.5 Compaction of the Error Dictionary

In this section, we describe two algorithms to compact actuation matrices of synthesis results for bioassay. Corresponding algorithms for de-compaction of error dictionary also will be discussed.

3.5.1 Compaction of the Actuation Matrix

Suppose the actuation matrix $\mathscr{A}$ is an $(M \times N)$-by-T sparse matrix with $\mathscr{K}$ non-zero elements, where M and N refer to the array dimensions and T is the number of clock circles for the bioassay. We describe two algorithms for the compaction of the actuation matrix. The compaction ratio is defined as the number of elements before compaction divided by the number of elements in the matrix after compaction.

Method I: COO (coordinate list) compaction [26]. $\mathscr{A}$ can be stored as $\mathscr{N}$ three-tuples. Each tuple records the row indices, column indices and numerical values of a non-zero element in the matrix. Thus the number of elements in the actuation matrix is compacted from NMT to $3\mathscr{K}$, and its compaction ratio R_I can be calculated as $NMT/3\mathscr{K}$. For example, the matrix M_{Mix_1} in Sect. 3.3 has 36 elements, and seven of them are non-zero elements. Thus M_{Mix_1} can be compacted to 7 three-tuples, which have 21 elements in total: $(1, 1, v_1)$, $(2, 2, v_1)$, $\ldots$, $(6, 4, v_2)$, $(6, 6, v_2)$. The compaction ratio for matrix M_{Mix_1} is equal to 1.71.

Method II: Run-length encoding [27]. The actuation matrix of each dictionary entry can be reformatted to a signal sequence. The actuation sequence in which the same status value occurs in consecutive clock cycles can be compacted according to the single status value and the corresponding count number. For the actuation sequences for mixing and dispensing operations, we observe that "0" often occurs consecutively. Thus run-length encoding is applied to the "all-zero segments", i.e., runs of 0s in the actuation matrix. In the compaction result, non-zero elements are represented by their values, and runs of 0s are represented by the corresponding count number. For example, the compacted vector "$X_v a Y_v b$" stands for the following sequence:

$$X_v \underbrace{0 \dots 00}_{a} Y_v \underbrace{0 \dots 00}_{b}.$$

The $\mathcal{K}$ non-zero elements will "divide" the MN column vectors in $\mathcal{A}$ into at most $MN + \mathcal{K}$ all-zero segments, and each of these segments is represented by its number of zeros. Thus the actuation matrix can be compacted to a vector with $MN + 2\mathcal{K}$ elements, and its compaction ratio can be derived as $MNT/(MN + 2\mathcal{K})$. For the matrix M_{Mix_1} shown in Sect. 3.3, its first column is divided into two segments, where the first segment is the single non-zero element "v_1" and the second segment is five consecutive "0", so the first column can be compacted to "$(v_1)(5)$". Similarly, the second column is divided into three segments, and it is compacted to "$(1)(v_1)(4)$". The compaction ratio for matrix M_{Mix_1} is equal to 1.80.

Based on the above discussion, the relative performance of Method I and Method II depends on the relationship between MN and $\mathcal{K}$. When $\mathcal{K} > MN$, Method II offer higher compaction than Method I; otherwise, Method I is more efficient. Both compaction algorithms can be implemented when generating the error dictionary.

3.5.2 De-Compaction of the Error Dictionary

Before the execution of the bioassay, the error dictionary must be compacted and stored in the memory of the FPGA. Both compaction Method I and Method II discussed above are lossless and reversible. When an error occurs on the biochip is reported by the on-chip sensors, the controller accesses corresponding compacted re-synthesis outcome and restores it to the complete actuation matrix with the appropriate electrode actuation vectors.

The de-compaction procedure can be performed by de-compaction modules implemented on an FPGA. As explained above, by applying compaction Method I, the $(M \times N)$-by-T actuation matrix is compacted into a $1 \times \mathcal{K}$ vector. Each element in the vector represents a non-zero element in the original uncompacted matrix, denoted by (i, j, v_{ij}), where i and j are the row and column indices of the non-zero element; v_{ij} is the non-zero value.

Since at each time moment, we only need to apply actuation signals on $M \times N$ electrodes, the actuation matrix is de-compacted "segment by segment" during the de-compaction procedure. Here we assume that the memory space required for each element in the original actuation matrix is n_b bits. In the de-compaction module, the de-compacted $M \times N$ actuation sub-matrix will be stored in an $(M \times N \times n_b)$-bit register. Before applying the actuation matrix on the electrodes of the biochip, the data stored in the $(M \times N \times n_b)$-bit register is reset to zero. Next the de-compaction module "fills" non-zero elements into the $(M \times N \times n_b)$-bit register, i.e., the module assigns the elements in the $1 \times \mathcal{K}$ vector to the $(M \times N \times n_b)$-bit register. When the de-compaction procedure of one $M \times N$ actuation sub-matrix is finished, the corresponding actuation signals are applied to electrodes on the biochip. The data stored in the $(M \times N \times n_b)$-bit register will be reset to zero again, and the next segment in the compacted actuation matrix will be de-compacted.

The compaction results derived from Method II can be restored to the original actuation matrix in a similar way.

The clock frequency of the biochip is usually between 1 and 100 Hz, i.e., moving a droplet from one electrode to the adjacent electrode takes 10 ms to 1 s [25]. On the other hand, clock frequency for de-compaction and writing data into the $(M \times N \times n_b)$-bit register can be as high as 16 MHz [17]. The frequency of the FPGA is several orders of magnitudes higher than the frequency of fluid-handling operations, hence the total time needed for accessing the dictionary, de-compaction, and transfer of data from the FPGA to buffers in the peripheral circuit is negligible compared to the operation time of the biochip.

3.6 Implementation of Dictionary-Based Error Recovery on FPGA

The circuitry for error recovery on a cyberphysical microfluidic biochip consists of four main modules, i.e., (1) the sensing module for the detection of errors, (2) the memory module for the storage of the error dictionary, (3) the FSM module for the dynamic adjustment of actuation sequences when errors occur, and (4) the de-compaction module for decoding the actuation matrices. All four modules can be implemented on an FPGA.

The modules are described using Verilog, and synthesized using Quartus II [28]. All functional and timing simulations for the modules are performed using ModelSim-Altera [29]. The FPGA used in the simulation belongs to the family of Cyclone IV [30], which includes a series of devices. The maximum numbers of I/O ports and logic elements provided by these devices are different [30]. While synthesizing the modules, Quartus II automatically selects the suitable devices based on the required number of I/O ports and logic elements. The maximum on-chip memory that the family of Cyclone IV can provide is 0.83 MB.

The interconnection of the modules is presented in Fig. 3.5, and the detailed implementation of these modules in an FPGA is discussed below.

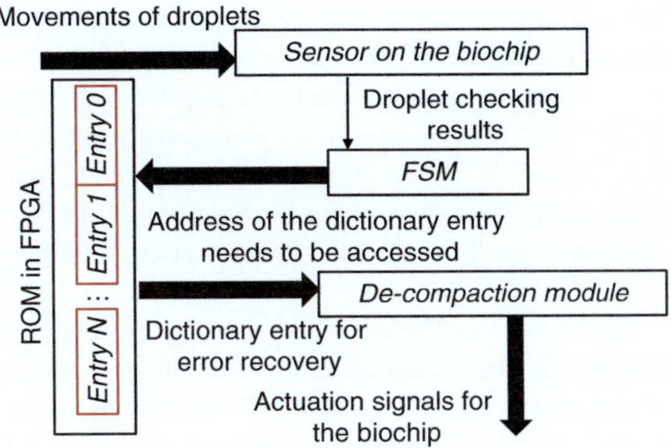

Fig. 3.5 Modules and their interconnections for the dictionary-based error recovery system

3.6.1   Sensing Module

The sensing module can be designed on a biochip based on different sensing techniques. Here, we use the imaging-based droplet detection method as the example to illustrate the implementation of sensing modules in FPGA. Note that the error-recovery algorithm approach proposed in this chapter is general and it can be applied to cyberphysical biochips with various sensing systems.

In Sect. 2.2.1, an imaging-based droplet-tracking algorithm is discussed. During the execution of the bioassay, an image sensor is used to monitor the entire biochip. Based on the images captured by the image sensor, the control software can automatically search for droplets by "template matching" (Sect. 2.2.1). The acquired images are first converted to two-dimensional matrices. For example, an image with $M_w \times N_w$ pixels will be converted to an $M_w \times N_w$ matrix, in which each element represents one pixel of the image.

An image of a typical droplet on the biochip is selected as the "template" (which is written as T). Each time, the software crops a sub-image (T_s) from the image of the entire biochip. The cropped sub-image and the template image are considered as two vectors, and the correlation index of these two vectors represent the similarity between the template and the cropped sub-image. In order to locate the positions of droplets on the biochip precisely, the correlation index is calculated on a pixel-by-pixel basis [14]. Therefore, if the image for the entire biochip has $M_w \times N_w$ pixels and the template image has $M_s \times N_s$ pixels, the calculation of the correlation factor between the template image and the cropped sub-image will be implemented $(M_w - M_s) \times (N_w - N_s)$ times. The computational complexity of droplet tracking is significantly high when there is a large number of pixels in the image of the entire biochip, hence the image-based droplet-tracking method need to be performed by the software on a computer [14].

Based on the characteristics of the digital microfluidic biochip, the image-based droplet-tracking procedure can be simplified and accelerated. Hence all the calculations can be performed on the FPGA. The movements of droplets on the biochip are synchronized under clock control. All the droplets are expected to stay at the center of their corresponding electrodes at the beginning of each clock cycle. Then, actuation signals are applied to all electrodes at the same time, and the droplets are moved towards their destination electrodes concurrently. At the end of the clock cycle, all the droplets are expected to stay at the center of their destination electrodes. Therefore, in order to monitor the movement of droplets, we need to check the position of each droplet at the beginning of each clock cycle for only once.

Since the biochip consists of a discrete electrode array, the image of the entire biochip can be partitioned into sub-images, and each sub-image shows the status of one electrode. The template T is selected as the image of an electrode on which there is a typical droplet. Assume that there are K electrodes on the biochip, the sub-images derived can be written as I_1, I_2, $\ldots$, I_K. By comparing T with these K sub-images, the positions of the droplets can be determined. Therefore, at each clock cycle, the calculation for the correlation index between the template and the sub-image will be implemented only K times.

The calculation of the correlation index between two vectors can be implemented through the image-based droplet-tracking module. In our simulation, we use the picture of a biochip with 15 electrodes to test the functionality of the module, and each sub-image contains 27×21 pixels. When the working frequency of the FPGA is 20 MHz, the time spent on checking all 15 sub-images is 0.45 ms, while the time to move a droplet from one electrode to another adjacent electrode is usually between 10 ms and 1 s [25]. The time spent on checking the droplet is negligible when compared with the operation time of the biochip. Hence, the imaging-based droplet-tracking module implemented on FPGA can be used as a real-time sensor for the biochip.

3.6.2 Memory for Storage of the Error Dictionary

The error dictionary of the bioassay is stored in the read-only memory (ROM) of the FPGA. During the fault simulation of the bioassay, entries of the error dictionary are generated, compacted, and then written into a "memory initialization file" (MIF). The MIF specifies the content for each memory cell in ROM. At the same time, a table that records the starting and ending addresses of each dictionary entry is generated and loaded into the module of the FSM (Fig. 3.5).

After the compilation of the FPGA project, the contents of the ROM are initialized by the MIF. When the FSM sends the addresses of the memory cells to the ROM, the data stored in corresponding cells are sent to the output of the ROM.

3.6.3 FSM Module

As introduced in Sect. 3.1, automatic error recovery can be implemented by an FSM. The input of the FSM is the detection results from sensing module. The outputs of the FSM are the starting and ending storage addresses of the dictionary entry that must be applied to the biochip. When the detection module sends a signal indicating that an error has occurred, the FSM transits from one state to another state for error recovery. The starting and ending storage addresses of the dictionary entry that will be applied to the biochip are changed accordingly.

The output of the FSM is connected to the inputs of the ROM. When the FSM sends the addresses of the dictionary entry, the data that are stored in the corresponding memory cells are sent to the output of the ROM. As the data stored in the ROM are the compacted dictionary entries, these data must be de-compacted before that can be applied to the electrodes on the biochip. Therefore, the output of the ROM is connected to the input of the de-compaction module, as shown in Fig. 3.5.

3.6.4 De-Compaction Module

The working principle of the de-compaction module is introduced in Sect. 3.5.2. The input of the de-compaction module is the compacted actuation sequence, and the output is the de-compacted actuation signal sequence that can be directly applied on the biochip.

The de-compaction modules that correspond to compaction Method I and Method II are both implemented by the FPGA. When the working frequency of the FPGA is set as 20 MHz, one dictionary entry with 930 elements, where each element is an 8-bit binary number, can be decoded in less than 50 μs. Therefore, the time spent on dictionary de-compaction is negligible, when compared with the time required to move a droplet from one electrode to an adjacent electrode.

3.6.5 Resource Report for Synthesized Modules

The modules are synthesized by Quartus II. The resource reports for the synthesized droplet-tracking module and the de-compaction module are listed in Table 3.4. The FPGA resource occupied by the synthesized controller module depends on the total number of states in the FSM. The relationship between the number of states and the number of logic elements is shown in Table 3.5. In the simulation, the actual FPGA device selected by Quartus II is EP4CGX15BF14C6.

Table 3.4 Resource report for synthesized modules

	# Logic elements	# Combinational functions	# Logic registers	# Total registers	# Total pins
Image-based droplet tracking	4,631	4,627	230	230	22
De-compaction (Method I)	249	249	144	14	42
De-compaction (Method II)	42	42	31	31	42

Table 3.5 Resource report for synthesized FSMs with different number of states

No. state of the FSM	# Logic elements	# Combinational functions	# Logic registers	# Total registers	# Total pins
50	62	35	50	50	13
500	390	390	9	9	13
800	680	680	10	10	18
1,200	419	419	11	11	18
1,600	168	168	11	11	35

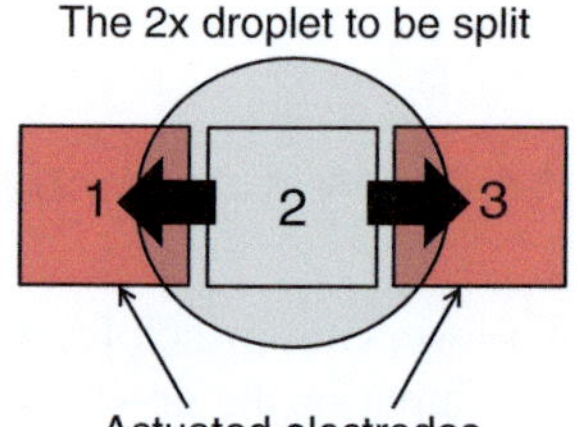

Fig. 3.6 Splitting of a droplet: both Electrodes *1* and *3* are actuated, and electrowetting forces are generated on the surfaces of these two electrodes

3.7 Fault Simulation in the Presence of Chip-Parameter Variations

In this section, we examine the nature of the errors that occur in splitting operations and present a parameter variation-aware fault simulation algorithm. By considering the variations in the parameters of the biochip electrodes, the simulator can mimic the erroneous behavior during the execution of bioassay.

As introduced in Sect. 3.1, error occurrence in the biochip is defined as the occurrences of droplets with abnormal volumes or concentrations. Experimental results with fabricated biochips show that the generation of abnormal droplets in asymmetric splitting operations is the primary cause of errors that occur in biochips [24]. Furthermore, as introduced in Sect. 3.2.1, splitting operations are frequently implemented in bioassays. Therefore, most of the errors that occur during the execution of the bioassay are caused by the asymmetric splitting operations [31]. As shown in Fig. 3.6, to split the droplet on Electrode 2, Electrode 1 and Electrode 3 are actuated at the same time. Two electrowetting forces, represented as F_1 and F_3, are generated on the surfaces of Electrodes 1 and 3, respectively. The droplet on Electrode 2 will be split under F_1 and F_3, which are applied in opposite directions.

The electrowetting force applied on a droplet can be written as [32]:

$$F = \left(\frac{\varepsilon_{rd}\varepsilon_{rh}}{t_h\varepsilon_{rd} + t_d\varepsilon_{rh}}\right)\frac{\varepsilon_0 V^2}{2}\frac{dA_p}{dx}, \tag{3.1}$$

where ε_{rd} is the dielectric constant of dielectric layer, ε_{rh} is the dielectric constant of the hydrophobic layer, t_d is the thickness of the insulator layer, t_h is the thickness of the hydrophobic layer, ε_0 is the permittivity of vacuum, x is the position of the droplet, V is the voltage applied to the actuated electrode, and A_p is the area of the overlapping region between the droplet and the actuated electrode [32].

In an ideal situation, at the beginning of the splitting operation, the droplet in Fig. 3.6 stays at the center of Electrode 2. The overlap region between the droplet and Electrode 1 (A_{p_1}) are the same with the overlap region between the droplet and Electrode 3 (A_{p_3}), therefore we have: $\frac{dA_{p_1}}{dx} = \frac{dA_{p_3}}{dx}$. For Electrodes 1 and 3, their parameters t_d and t_h are also the same. Thus the forces generated on the surfaces of Electrodes 1 and 3 are symmetric, i.e., $\frac{\|F_1\|}{\|F_3\|} = 1$. The droplet will be split over the two electrodes into droplets of equal size.

However, due to the randomness inherent in the fabrication process, the parameters t_h and t_d may vary from electrode to electrode, and the electrowetting forces generated on the surfaces of Electrodes 1 and 3 may not be exactly the same. The droplet will be split under asymmetric forces in this case.

The droplet on Electrode 2 may be split into two droplets that have unequal volumes if asymmetric forces are applied [32]. For example, when we split a droplet whose volume is 2 units, if $\frac{\|F_1\|}{\|F_3\|} = 1.10$, the volumes of the droplets derived will be 1.13 and 0.87 units, respectively [32]. If the "standard" volume of a droplet is 1 unit, and the error limit for the volume of droplets is 10 %, then both of these droplets will be abnormal. In this case, errors are generated on the biochip.

The simulator can mimic the behavior of splitting operations in the bioassay by considering the distributions of parameters t_d and t_h for the biochip. For any electrode E_i, the thickness of the insulator layer is written as $t_d(E_i)$, the thickness of the hydrophobic layer is written as $t_h(E_i)$. For given distributions of parameters t_d and t_h for the biochip, and the given synthesis result of the bioassay, the simulator can calculate the ratio of the two electrowetting forces generated in each splitting operation.

We assume that the splitting operation O_S is performed by electrodes E_L and E_R. According to (3.1), the ratio of the electrowetting forces generated on E_L and E_R (written as F_{E_L} and F_{E_R}) can be expressed as:

$$\frac{\|F_{E_R}\|}{\|F_{E_L}\|} = \frac{(t_h(E_L)\varepsilon_{rd} + t_d(E_L)\varepsilon_{rh})}{(t_h(E_R)\varepsilon_{rd} + t_d(E_R)\varepsilon_{rh})}. \tag{3.2}$$

If the ratio of $\|F_{E_R}\|$ to $\|F_{E_L}\|$ is out of acceptable range, the splitting operation O_S will generate two droplets that have abnormal volumes. In this way, an error is generated during fault simulation, i.e., we mimic the occurrences of errors by running the fault simulation under parameter variations.

3.8 Simulation Results

In this section, we first proof that the actuation matrices for bioassays are sparse. Next, we present simulation results for four widely used laboratory protocols, namely exponential dilution of a protein sample, interpolation dilution of a protein sample, mixing tree bioassay, and PCR bioassay, respectively [16, 23]. Then the simulation result of a new benchmark for a flash chemistry application is presented [33].

The error dictionaries for these four bioassays are generated, and the compaction algorithms for error dictionaries are applied. Fault simulation to mimic the occurrences of errors are also carried out.

3.8.1 Exponential Dilution of a Protein Sample

3.8.1.1 Generating Error Dictionaries

The exponential dilution bioassay contains 103 operations. The sequencing graph is shown in Fig. 2.20b and the detailed description of the bioassay can be found in [16].

By applying the PRSA-based synthesis algorithm [16], the synthesis result for the bioassay can be derived. When running the bioassay on a 10×10 electrode array and no erroneous operation occurs, the completion time is 208 s. Based on the synthesis result for the error-free case, the dynamic re-synthesis algorithm proposed in [35] is applied to generate the error dictionary.

In order to generate entries of the error dictionary, erroneous operations are inserted into the bioassay. By considering each operation as a possible erroneous operation, 103 dictionary entries with single error that occurs in the bioassay are generated. If we set the upper limit for the number of erroneous operations to be two, the number of possible combinations of errors that must be considered is $\binom{103}{2}$, i.e., $(103 \times 102)/2$ (here we assume that errors will not occur in error recovery operations). Therefore, the total number of entries $\mathcal{N}_e$ in the dictionary is given by: $\mathcal{N}_e = 103 + (103 \times 102)/2 = 5356$.

The fault simulation is run on a 2.30 GHz Intel i3 dual-core processor with 8 GB of memory. The CPU time is 1,925.7 s for generating the error dictionary. The histogram for the numbers of extra droplets consumed in these 5,356 dictionary entries is shown in Fig. 3.7a. The maximum number of additional droplets consumed in the process of error-recovery is 8.

When we consider the cost of error recovery, not all these 5,356 dictionary entries will be selected as "effective entries" in the error dictionary. For some biochemistry experiments, the cost of precious samples and reagents can be greater than the cost of the biochips. If "too many" droplets are consumed in recovering from the error, running the bioassay on a new biochip can be more cost-effective. In this situation, the corresponding synthesis results will not be recorded in the error dictionary.

Fig. 3.7 (**a**) The histogram for the number of extra droplets consumed in all possible situations of exponential dilution bioassay; (**b**) the histogram for the time spans of effective dictionary entries when $N_{max} = 5$

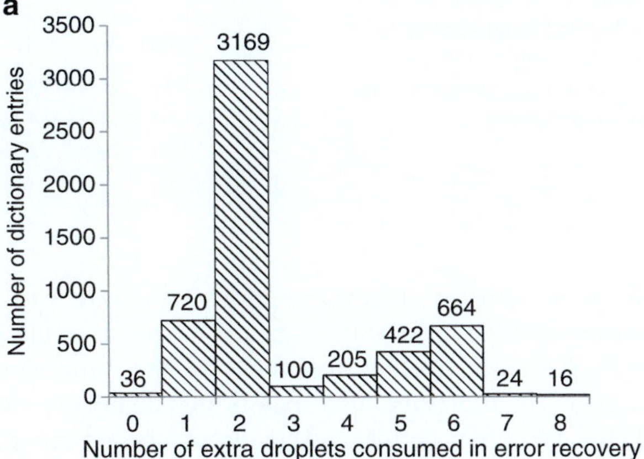

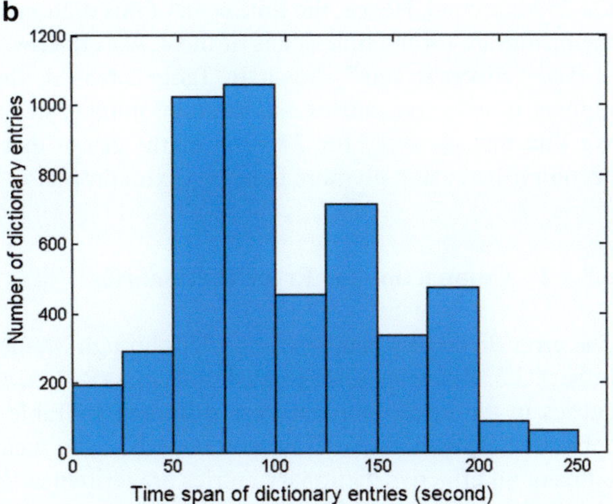

Table 3.6 Effective dictionary entries corresponding to different values of N_{max} in the exponential dilution bioassay

N_{max}	No. effective entries	Coverage rate (%)	Max. time span (s)	Min. time span (s)
2	4,013	74.92	222	5
3	4,113	76.79	230	5
5	4,747	88.63	237	5
8	5,356	100.00	237	5

The parameter N_{max} denotes the maximum number of additional droplets that can be used in error-recovery when we consider the cost of samples and reagents. When N_{max} is set as 2, 3, 5, and 8, the numbers of effective entries that are recorded in the dictionary are 4,013, 4,113, 4,747, and 5,356, respectively. The minimum and maximum time spans of these effective entries corresponding to different value of N_{max} can be found in Table 3.6. For example, when $N_{max} = 5$, the time spans of

Table 3.7 Compaction ratios of Method I and Method II corresponding to different values of N_{max} in exponential dilution bioassay

N_{max}	S_O (MB)	$\overline{R}_I$	σ_I	S_I (MB)	$\overline{R}_{II}$	σ_{II}	S_{II} (MB)
2	21.12	24.13	4.43	0.90	34.89	6.12	0.62
3	21.61	24.23	4.53	0.93	34.98	6.26	0.64
5	24.77	24.09	4.45	1.12	34.77	6.11	0.77
8	28.75	23.82	4.25	1.40	34.47	5.89	0.96

the 4,747 effective dictionary entries vary from 5 to 237 s. One of the dictionary entries with the shortest time span corresponds to the error that occurs at the 208th second; after recovering from the error, the entire bioassay is finished at the 213th second. Therefore, the time span of this dictionary entry is 5 s. One of the longest dictionary entries corresponds to two errors that occur at the 15th and 22nd second, respectively; after recovering from these two errors, the entire bioassay finishes at the 259th second. Hence, the time span of this dictionary is 237 s. Figure 3.7b shows the histogram for the time spans of the 4,747 effective dictionary entries.

The "coverage rate" shown in Table 3.6 is defined as the ratio between the number of effective entries and the total number of entries in the error dictionary. We find that recovery for 74.92 % of the errors that occur in a bioassay can be accomplished using no more than two extra droplets.

3.8.1.2 Compaction for Error Dictionaries

The error dictionary can be compacted through Method I and Method II discussed in Sect. 3.5. When we set the value of N_{max} as 2, 3, 5, and 8, the number of effective entries in the error dictionary are different. In Table 3.7, the original size of the "effective error dictionary" is written as S_O; the mean values for the compaction ratios of all effective dictionary entries are written as $\overline{R}_I$ (derived by Method I) and $\overline{R}_{II}$ (derived by Method II), respectively; the standard deviations for the compaction ratio of all effective dictionary entries are written as σ_I (derived by Method I) and σ_{II} (derived by Method II), respectively. Finally, the sizes of the dictionary after compaction are written as S_I (derived by Method I) and S_{II} (derived by Method II), respectively.

From Table 3.7, we can find that the total memory required to store the original effective error dictionary is 21.12~28.75 MB, which is much larger than memory available on a low-cost FPGA. After compaction, the final size of the compacted dictionary can be smaller than 1 MB.

3.8.1.3 Fault Simulation Results

After the generation and compaction of the error dictionary, next we run fault simulation for the exponential dilution bioassay.

As discussed in Sect. 3.7, if the ratio of electrowetting forces are out of the acceptable range, abnormal droplets will be generated. It is important to note that, in the bioassay protocols, there is an underlying assumption that the splitting of a droplet is the last step in the mixing/dilution procedure, i.e., splitting operations are executed at the end of each mixing/dilution operation. If droplets with abnormal volumes are generated when splitting the droplet at the end of operation $O_{M/D}$, this situation is defined as "an error occurs in $O_{M/D}$".

In the fault simulation setup considered here, we set the distribution function of the Gaussian random variable $R_c = (t_h \varepsilon_{rd} + t_d \varepsilon_{rh})$ as $r_c \cdot N(1, 0.03^2)$, where r_c is a nominal value, and $N(1, 0.03^2)$ is the Gaussian distribution function with mean 1 and variance 0.03^2.

For a typical biochip, $t_d \varepsilon_{rh} \gg t_h \varepsilon_{rd}$ for R_c [8]; the distribution function for R_c is estimated according to the process variance of t_d. For a typical biochip fabrication process, the average spread in t_d is 11 nm [37], while the value of t_d usually is $450 \sim 800$ nm [8, 37]. Therefore, it is reasonable to set the normalized variance for R_d as 3 %.

When a droplet is split by electrowetting forces F_1 and F_2, and $\frac{\|F_1\|}{\|F_2\|} \geq 1.10$ (alternatively, $\frac{\|F_2\|}{\|F_1\|} \geq 1.10$), we assume that an error occurs. We run the fault simulation procedure a total of 1,000 times. Among the simulation results for the 1,000 runs, one error occurs 233 times and none of these 1,000 runs include more than one error.

For the 233 runs with errors, the distributions for the number of extra droplets consumed in error recovery and the total completion time of bioassay derived by the proposed error recovery method are shown in Fig. 3.8a, b, respectively. As shown in Fig. 3.8, for these 233 runs, the maximum number of droplets consumed in error recovery is 5; the maximum total completion time for the bioassay is 237 s.

Part of the simulation results for the 233 runs with errors can be found in Table 3.8. After the error is detected, the biochip can derive the re-synthesis results either by the Online Synthesis Strategy (ONS) [12], Fast Online Synthesis [36], or by looking up the prepared error dictionary. The response time in Table 3.8 is defined as the time spent in deriving the actuation sequences to recover the error. Here we assume that both ONS and Fast Online Synthesis are using the hardware setup presented in Fig. 2.5, i.e., the biochip and the computer are connected via an FPGA/signle board controller. The response times of ONS and Fast Online Synthesis include the time spent on writing the new actuation matrices into the memory initialization files of the FPGA (Table 3.8).

Note that during on-line re-synthesis, all fluid-handling operations for the bioassay are suspended. The execution time in Table 3.8 is calculated from the beginning to the end of the bioassay, without considering the CPU time spent in re-synthesis. Thus the total completion time for the bioassay is the sum of the response time and the execution time.

As shown in Table 3.8, the CPU time needed for performing on-line re-synthesis by both ONS and Fast Online Synthesis are longer than the proposed method. It is important to note that for the laboratorial experiments in biochemistry, the

Fig. 3.8 (a) The histogram for the numbers of extra droplets consumed in the 233 runs in which 233 errors are generated; (b) the histogram for the time spans of entries in the 233 runs in which errors are generated

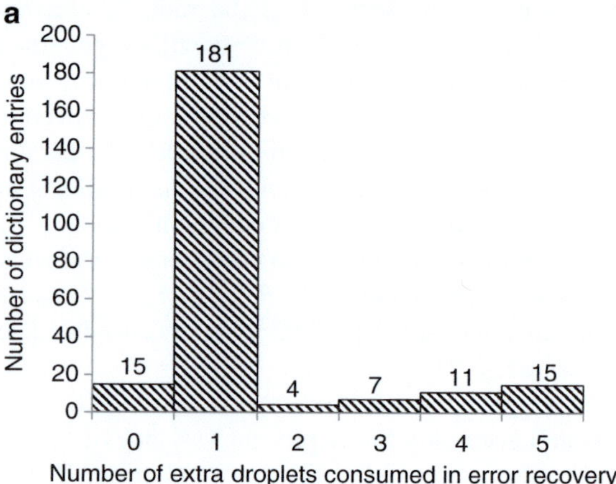

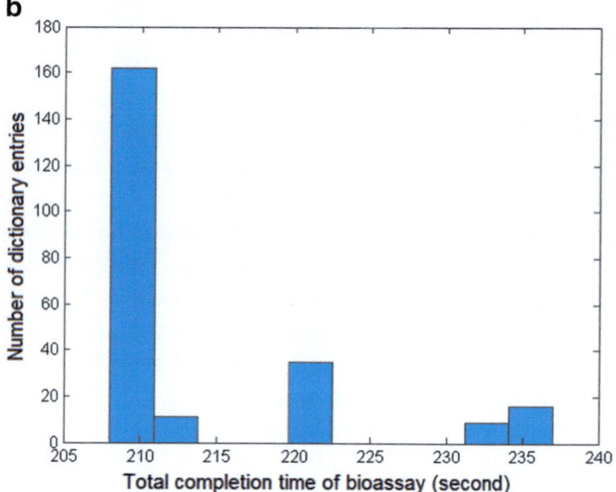

reaction time for each step in the chemical synthesis is often very short, and the intermediate products may degrade or decompose quickly [3, 33]. Thus the increase in response time introduced by the on-line calculations required in [16, 36] may have an adverse impact on reaction outcome [3, 33]. Therefore, the proposed dictionary-based method can recover from the error "seamlessly" and minimize the degeneration of intermediate products of the bioassay. It is important to note that, even though ONS and Fast Online Synthesis can achieve online re-synthesis, the calculation has to be performed on the computer. Therefore, they are not suitable for the low-cost cyberphysical system that has only one single board controller/FPGA.

Table 3.8 Comparison of response times and bioassay completion times

Bioassay	Errors inserted	ONS [12]			Fast online synthesis [36]			Proposed method		
		Response time (s)	Execution time (s)	Total time (s)	Response time (s)	Execution time (s)	Total time (s)	Response time (s)	Execution time (s)	Total time (s)
Exponential dilution	Dlt_{26}	0.21	222	222.21	0.22	240	240.22	~ 0	208	208
	Mix_3	0.13	220	220.13	0.31	233	233.31	~ 0	213	213
	Dlt_2	0.88	222	222.88	0.94	240	240.94	~ 0	216	216
	Dlt_{21}	0.69	222	222.69	0.74	233	233.74	~ 0	208	208
	Dlt_{27}	0.52	222	222.52	0.54	233	233.54	~ 0	208	208
Interpolation dilution bioassay	DsB_2 and Dlt_{18}	0.48	203	203.48	0.43	237	237.43	~ 0	182	182
	Dlt_{18} and Dlt_{19}	0.46	196	196.46	0.45	230	230.45	~ 0	189	189
	Dlt_2 and Dlt_{29}	0.27	196	196.27	0.30	230	230.30	~ 0	182	182
	Dlt_8 and DsB_3	0.62	203	203.62	0.61	237	237.61	~ 0	182	182
	Dlt_{16} and Dlt_{18}	0.43	196	196.43	0.43	230	230.43	~ 0	187	187

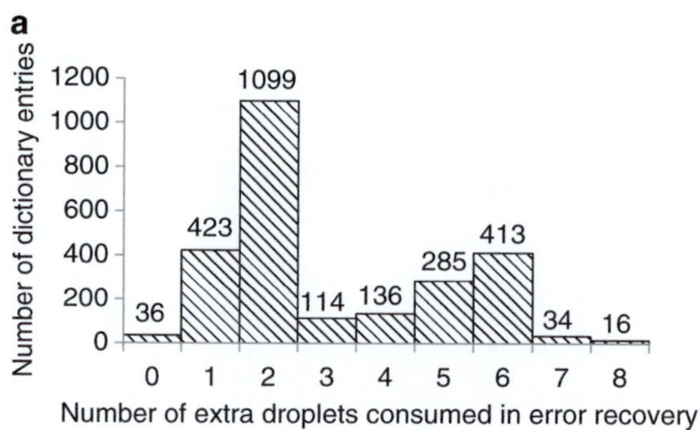

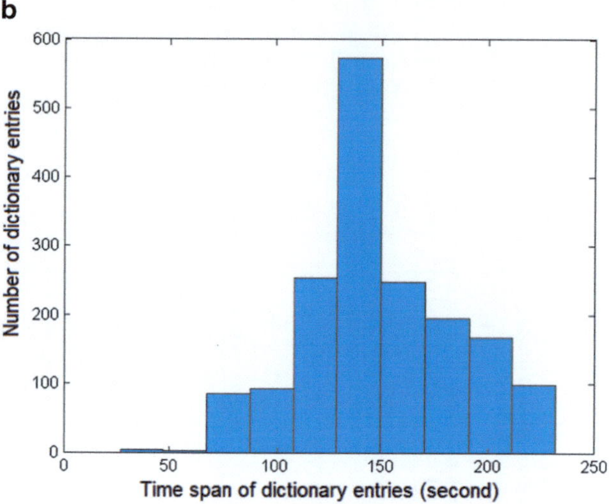

Fig. 3.9 (**a**) The histogram for the numbers of extra droplets consumed in all possible situations of the interpolation-based dilution bioassay; (**b**) the histogram for the time spans of entries in the error dictionary for the interpolation-based dilution bioassay (when $N_{max} = 3$)

3.8.2 *Interpolation Dilution of a Protein Sample*

The sequencing graph of interpolation-based dilution bioassay can be found in Fig. 2.20a [16]. There are 71 operations in the interpolation-based dilution bioassay. When no error occurs, the execution time of the bioassay is 182 s, and 32 droplets are consumed in the bioassay. Assume that the number of errors occur in the bioassay is no more than 2, then the total number of cases need to be considered is $\mathcal{N}_e = 71 + (71 \times 70)/2 = 2556$. For these 2,556 dictionary entries, the histogram for number of droplets consumed in error recovery is shown in Fig. 3.9a.

By setting different value for N_{max}, the number of effective entries in the dictionary will be different. For example, when N_{max} is set as 3, 5, and 8, the numbers of effective entries in the dictionary are 1,727, 2,156, and 2,556,

Table 3.9 Compaction ratio of Method I and Method II corresponding to different values of N_{max} in interpolation-based dilution bioassay

N_{max}	S_O (MB)	$\overline{R}_I$	σ_I	S_I (MB)	$\overline{R}_{II}$	σ_{II}	S_{II} (MB)
3	14.29	20.51	0.97	0.84	29.83	1.53	0.58
5	17.51	20.74	1.05	0.94	30.19	1.67	0.64
8	21.72	20.77	1.03	1.02	30.27	1.64	0.70

respectively. When $N_{max} = 3$, the histogram for the time span of effective dictionary entries is shown in Fig. 3.9b.

The error dictionary can be compacted by Method I and Method II discussed in Sect. 3.5. When setting the value of N_{max} as 3, 5, and 8, the corresponding simulation results are presented in Table 3.9. The parameters of Table 3.7 are defined in Sect. 3.8.1.

In the realistic fault simulation setup considered here, we set the distribution function of R_c as $r_c \cdot N(1, 0.035^2)$. We run the parameter various-aware fault simulation for 1000 times. When a droplet is split by electrowetting forces F_1 and F_2, and $\frac{\|F_1\|}{\|F_2\|} \geq 1.10$ (or $\frac{\|F_2\|}{\|F_1\|} \geq 1.10$), then we assume that an error occurs.

Among the simulation results of the 1,000 runs, 426 runs have one error occurring, 129 runs have two errors, and none of the runs lead to more than two errors. The simulation results corresponding to five runs can be found in Table 3.8. We can find that the response time and the total completion time of the bioassay are reduced using the proposed dictionary-based re-synthesis procedure.

3.8.3 *Mixing Tree Bioassay*

In this part, we present the simulation results for a biochemical benchmark designed in [23]. The biochemical benchmark is a solution preparation procedure, which has seven different kinds of reagent. The sequencing graph of the benchmark can be found in Fig. 3.10 [23]. The final output is the droplet which has the mixing ratio of $2 : 3 : 5 : 7 : 11 : 13 : 87$ [23]. The bioassay has 37 operations, and the number of droplets consumed is 10. When the bioassay runs on an 8×8 array, the completion time of the bioassay is 60 s. For all the 740 dictionary entries, the numbers of droplets consumed in error recovery are no more than 4. This is because in the protocol of mixing tree bioassay, each mixing/dilution operation generates a redundant copy droplet. These redundant droplets are stored on the biochip and they can be utilized during error-recovery process. Hence only a small number of extra droplets need to be dispensed for recovering the errors.

For compaction Method I, the compaction ratios for the dictionary entries vary from 2.1 to 13.4. The average compaction ratio for these entries is 4.8 and the standard deviation is 3.2. For compaction Method II, the compaction ratios for the dictionary entries vary from 2.8 to 18.2. The average compaction ratio for these entries is 6.6 and the standard deviation is 4.4. The size of the error dictionary can be reduced from 6.15 to 1.18 MB (Method I) and 0.87 MB (Method II).

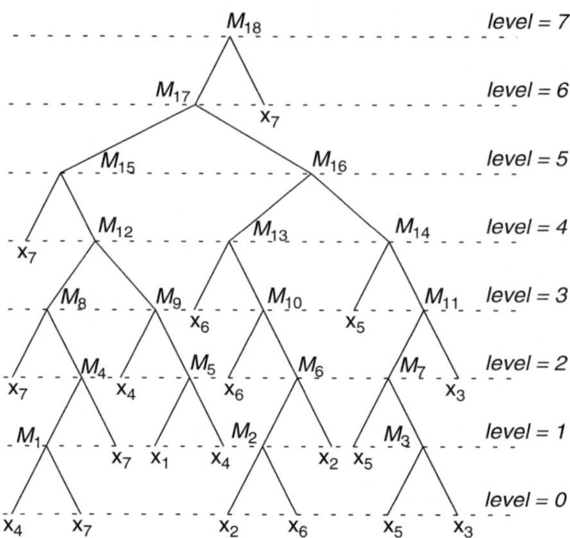

Fig. 3.10 Sequencing graph for a solution preparation procedure. The inputs of the mixing procedure are seven different kinds of samples/reagents and the final output is the droplet that has the mixing ratio of 2:3:5:7:11:13:87 [23]. These seven samples/reagents are loaded into reservoirs $R_{1\sim7}$ on the PCR biochip

Table 3.10 Compaction ratios of actuation matrices of PCR assay

Errors inserted	Bioassay time (s)	No. elements in the dictionary entry	No. non-zero elements in the dictionary entry	R_I	R_{II}
M_1	24	1,920	56	11.43	10.00
M_2 and M_7	34	2,720	66	13.74	12.83
M_6	27	2,160	59	12.2	10.91
M_4 and M_5	24	1,920	59	10.85	9.70

3.8.4   PCR Bioassay

The description of the PCR bioassay can be found in [16]. In the error-free case, the completion time is 18 s, and the size of the biochip is 8×8. We randomly generate 20 error cases, and part of the simulation results are shown in Table 3.10. We find that for this small bioassay with a small number of operations, Method I is more effective than Method II.

Table 3.11 Compaction ratios of actuation matrices of an assay in flash chemistry

Errors inserted	Bioassay time (s)	No. elements in the dictionary entry	No. non-zero elements in the dictionary entry	R_I	R_{II}
M_2 in $\mathbb{S}_1$	0.32	3,200	93	11.47	11.19
M_4 in $\mathbb{S}_1$	0.25	2,500	86	9.69	9.19
M_1 in $\mathbb{S}_2$	0.19	1,900	94	6.73	6.60
M_2 in $\mathbb{S}_2$	0.19	1,900	87	7.28	6.93

3.8.5 Flash Chemistry

As introduced in [33], a microfluidic biochip is one of the promising hardware platforms for flash chemistry. In this subsection, the simulation results for two representative assays of flash chemistry are presented. These two assays are "synthesis of unsymmetrical diarylethenes" ($\mathbb{S}_1$) and the "bromine-lithium exchange reaction of o-dibromobenzene" ($\mathbb{S}_2$) [33]. The sequence graphs for $\mathbb{S}_1$ and $\mathbb{S}_2$ can be found in Fig. 3.11 [33]. We assume that the two assays are concurrently executed on a 10×10 electrode array.

It is important to note that the "microtube reactors" shown in Fig. 3.11 can be viewed as time controllers of reactions in flash chemistry. Since the reaction time can be easily controlled on a digital microfluidic biochip (for example, we can merge two reagent droplets together at a pre-determined time moment to start the reaction, and separate the droplets at a given time moment to end the reaction), in the synthesis procedure, we consider "reactors" as the same module for mixing operation, and not assign specific "reactors" on the biochip. Therefore, the assay "synthesis of unsymmetrical diarylethenes" includes four mixing operations and five dispensing operations for five different kinds of reagents/samples; the assay "bromine-lithium exchange reaction of o-dibromobenzene" includes two mixing operations and three dispensing operations for three different kinds of reagents/samples.

The reaction times in flash chemistry usually are very short. Here we assume that the working frequency of the biochip is 100 Hz, and each mixing operation on a 2×4 array takes 40 ms [33]. Since dispensing each reagent/sample droplet takes 4 clock cycles [24], the completion time for each dispensing operation is set as 40 ms. In the error-free case, the completion time of these three bioassay is 190 ms.

As the total number of operations in the "combined bioassay" is 14, the number of dictionary entries we generated is $\mathcal{N}_e = 14 + (14 \times 13)/2 = 105$. The average response time for the proposed method is nearly zero. We have also derived the compaction ratios for all the dictionary entries. Part of the simulation results can be found in Table 3.11. For this small assay, Method I is more effective than Method II.

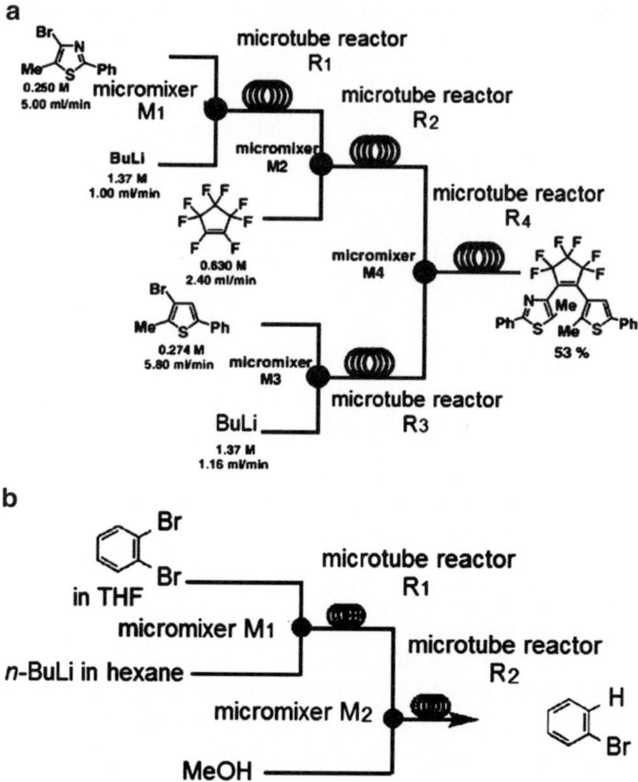

Fig. 3.11 The sequencing graphs for two representative assays in flash chemistry: (**a**) synthesis of unsymmetrical diarylethenes; (**b**) bromine-lithium exchange reaction of o-dibromobenzene [33]

3.9   Chapter Summary and Conclusions

In this chapter, we have described a dictionary-based hardware-assisted error recovery approach for flash chemistry based on digital microfluidic biochips. The proposed method minimizes the response time for error recovery, and provides precise control of on-chip experiments for chemical synthesis. The proposed error-recovery procedure and dynamic re-synthesis of a reaction can be implemented in real-time on an FPGA. In order to store the error dictionary in the limited memory available in the FPGA, we have presented two compaction techniques for reducing the size of the error dictionary. We have used five laboratorial protocols to show that, compared to software-based methods, the proposed dictionary-based error-recovery method has negligible impact on the time-to response. It also requires less complex experimental setup, and needs only a small amount of memory on the FPGA board.

The future work include: (i) exploring applications of cyberphysical microfluidics in cell sorting [38] and chip cooling [39]; (ii) identifying alternative methods (e.g., Huffman encoding [40]) for compacting the data in error dictionaries.

References

1. J. Yoshida, *Flash Chemistry: Fast Organic Synthesis in Microsystems*, Hoboken, NJ: Wiley, 2008.
2. T. Iwasakia, A. Nagakib, and J. Yoshida, "Microsystem controlled cationic polymerization of vinyl ethers initiated by CF_3SO_3H", *Chem. Commun.*, vol. 2007, no. 12, pp. 1263–1265, 2007.
3. J. Yoshida, "Flash chemistry: flow microreactor synthesis based on high-resolution reaction time control", *The Chemical Record*, vol 10, pp. 332–341, 2010.
4. T.-Y. Ho, K. Chakrabarty, and P. Pop, "Digital microfluidic biochips: Recent research and emerging challenges", *Proc. IEEE CODES+ISSS*, pp. 335–343, 2011.
5. R. Fair, A. Khlystov, T. Tailor, V. Ivanov, R. Evans, V. Srinivasan, V. Pamula, M. Pollack, P. Griffin, and J. Zhou, "Chemical and biological applications of digital-microfluidic devices", *IEEE Design & Test of Computers*, vol. 24, pp. 10–24, 2007.
6. E. Welch, Y.-Y. Lin, A. Madison, and R. Fair, "Picoliter DNA sequencing chemistry on an electrowetting-based digital microfluidic platform", *Biotech. J.*, vol. 6, pp. 165–176, 2011.
7. T. Xu, K. Chakrabarty, and V. K. Pamula, "Defect-tolerant design and optimization of a digital microfluidic biochip for protein crystallization", *IEEE Transactions on Computer-Aided Design of Integrated Circuits and Systems*, vol. 29, Issue 4, pp. 552–565, 2010.
8. Y.-Y. Lin, R. Evans, E. Welch, B.-N. Hsu, A. Madison, and R. Fair, "Low voltage electrowetting-on-dielectric platform using multi-layer insulators", *Sensors and Actuators, B: Chemical*, vol. 105, pp. 465–470, 2010.
9. T.-W. Huang, T.-Y. Ho, and K. Chakrabarty, "Reliability-oriented broadcast electrode-addressing for pin-constrained digital microfluidic biochips", *Proc. International Conference on Computer-Aided Design*, pp. 448–455, 2011.
10. P.-H. Yuh, C.-L. Yang, and Y.-W. Chang, "Placement of digital microfluidic biochips using the T-tree formulation", *Proc. IEEE/ACM Design Automation Conference*, pp. 931–934, 2006.
11. T. Xu and K. Chakrabarty, "Broadcast electrode-addressing for pin-constrained multi-functional digital microfluidic biochips", *Proc. IEEE/ACM Design Automation Conference*, pp. 173–178, 2008.
12. M. Alistar, P. Pop, and J. Madsen, "Online synthesis for error recovery in digital microfluidic biochips with operation variability", *Symposium on Design, Test, Integration and Packaging of MEMS/MOEMS*, pp. 53–58, 2012.
13. Y. Zhao, T. Xu, and K. Chakrabarty, "Broadcast electrode-addressing and scheduling methods for pin-constrained digital microfluidic biochips", *IEEE Transactions on Computer-Aided Design of Integrated Circuits and Systems*, vol. 30, Issue 7, pp. 986–999, 2011.
14. Y. Luo, K. Chakrabarty, and T.-Y. Ho, "Error recovery in cyberphysical digital-microfluidic biochips", *IEEE Transactions on Computer-Aided Design of Integrated Circuits and Systems*, vol. 32, Issue 1, pp. 59–72, 2013.
15. http://www.dvq.com/docs/brochures/intel_sbc_80_10.pdf (Intel Single Board Computer brochure), 1975.
16. F. Su and K. Chakrabarty, "High-level synthesis of digital microfluidic biochips", *ACM J. Emerging Tech. in Comp. Sys.*, vol. 3, January 2008.
17. http://www.altera.com/literature/hb/cyc/cyc_c51007.pdf (Cyclone Handbook Volume 1, Chapter 7: "On-chip memory implementations using Cyclone memory blocks")
18. http://www.altera.com/buy/buy-index.html (On-line purchase of Altera devices)

19. http://www.altera.com/products/devkits/altera/kit-cyc2-2C20N.html (On-line introduction and purchase of Cyclone II FPGA Starter Development Kit)
20. http://www.altera.com/literature/hb/nios2/edh_ed51008.pdf (Altera Handbook: "Memory system design")
21. S. Kirkpatrick, C. Gelatt, and M. Vecchi, "Optimization by simulated annealing", *Science*, vol. 220(4598), pp. 671–680, May 1983.
22. Y.-L. Hsieh, T.-Y. Ho, and K. Chakrabarty, "Design methodology for sample preparation on digital microfluidic biochips", *Proceedings of IEEE International Conference on Computer Design*, pp. 189–194, 2012.
23. S. Roy, B. Bhattacharya, P. Chakrabarti, and K. Chakrabarty, "Layout-aware solution preparation for biochemical analysis on a digital microfluidic biochip", *Proc. IEEE International Conference on VLSI Design*, pp. 171–176, 2011.
24. H. Ren, V. Srinivasan, and R. Fair, "Design and testing of an interpolating mixing architecture for electrowetting-based droplet-on-chip chemical dilution", *International Conference on Solid-State Sensors, Actuators and Microsystems*, pp. 619–622, 2003.
25. P. Paik, V. Pamula, and R. Fair, "Rapid droplet mixers for digital microfluidic systems", *Lab on a Chip*, vol. 3, pp. 253–259, 2003.
26. T. Bell and B. McKenzie, "Compression of sparse matrices by arithmetic coding", *Data Compression Conference*, pp. 23–32, 1998.
27. R. Wainwright and M. Sexton, "A study of sparse matrix representations for solving linear systems in a functional language", *Journal of Functional Programming*, Volume 2, Issue 1, pp. 61–72, 2012.
28. http://www.altera.com/products/software/quartus-ii/web-edition/qts-we-index.html
29. http://www.altera.com/products/software/quartus-ii/modelsim/qts-modelsim-index.html
30. http://www.altera.com/literature/hb/cyclone-iv/cyiv-51001.pdf (Datasheet of Cyclone IV FPGA Device Family)
31. H. Ren, R. Fair, and M. Pollack, "Automated on-chip droplet dispensing with volume control by electro-wetting actuation and capacitance metering", *Sensors and Actuators B*, Issue 98, pp. 319–327, 2004.
32. B. Bhattacharjee, *Study of Droplet Splitting in An Electrowetting based Digital Microfluidic System*, PhD Thesis, The University of British Columbia, Okanagan, CA, 2012.
33. J. Yoshida, *Flash Chemistry: Fast Organic Synthesis in Microsystems*, Hoboken, NJ: Wiley, 2008.
34. Y. Zhao, T. Xu, and K. Chakrabarty, "Integrated control-path design and error recovery in digital microfluidic lab-on-chip", *ACM JETC*, Vol. 6, No. 3, Article 11, 2010.
35. Y. Luo, K. Chakrabarty, and T.-Y. Ho, "A cyberphysical synthesis approach for error recovery in digital microfluidic biochips", *Proc. DATE*, pp. 1239–1244, 2012.
36. D. Grissom and P. Brisk, "Path scheduling on digital microfluidic biochips", *IEEE/ACM Design Automation Conference*, pp. 26–35, 2012.
37. H. Lee and J. Cho, "Development of conformal PDMS and Parylene coatings for micro-electronics and mems packaging", *Proc. International Mechanical Engineering Congress and Exposition*, pp. 1–5, 2005.
38. X. Huang, Q. Jia, M. Yan, H. Yu, and K. Yeo, "A super-resolution cmos imager for microfluidic imaging applications", *IEEE Biomedical Circuits and Systems Conference*, pp. 338–391, 2012.
39. H. Qian, C.-H. Chang, and H. Yu, "An efficient channel clustering and flow rate allocation algorithm for non-uniform microfluidic cooling of 3D integrated circuits", *Integration, the VLSI Journal*, vol. 46, no.1, pp. 57–68, 2013.
40. D. Huffman, "A method for the construction of minimum-redundancy codes", *Proceedings of the IRE*, Volume 40, Issue 9, pp. 1098–1101, 1952.

Chapter 4
Biochemistry Synthesis Under Completion-Time Uncertainties in Fluidic Operations

4.1 Introduction

Digital microfluidic biochips have emerged in recent years as a promising platform for implementing laboratory procedures in biochemistry [1, 2]. Biochemical assays, such as the dilution of samples and reagents, crystallization of protein molecules, on-chip chemistry for DNA sequencing, multiplexed real-time polymerase chain reaction (PCR), protein crystallization for drug discovery, and glucose measurement for blood serum, have been successfully implemented on such biochips.

The precision of fluidic operations is vital for accurately analyzing bioassays. For example, in the quantitative measurement for glucose concentration in blood [1], accurate measurements cannot be obtained if the mixing time for blood sample and enzymatic reagent is not precisely controlled. In order to determine appropriate settings for the completion-times of fluidic operations, bioassays need to be thoroughly characterized [3–5], i.e., fluidic operations must be repeatedly executed and monitored to obtain statistically significant results about their completion-times [3]. Based on these results, a module library is derived to define the completion-time for each type of operation, and this library is used as the guideline for the execution of on-chip operations. Table 4.1 shows an example of module library for dilution/mixing operations [6].

However, due to the inherent variability and randomness of biological/chemical processes [3–8], the problem of completion-time uncertainties in fluidic operations still remains after careful characterization of a bioassay. In practical applications, the completion-times of fluidic operations are random variables [3, 7, 8]. Therefore, an oversimplified module library that defines the completion-times of operations as constants cannot be used to precisely model the fluidic operations. If the module library is applied as a guideline for the execution of a bioassay, the bioassay yield may be low due to the uncertainties of completion-time in fluidic operations. Here the yield of bioassay is defined as the percentage of bioassay instances that can produce outcome droplets with expected concentrations.

© Springer International Publishing Switzerland 2015

Y. Luo et al., *Hardware/Software Co-Design and Optimization for Cyberphysical Integration in Digital Microfluidic Biochips*, DOI 10.1007/978-3-319-09006-1_4

Table 4.1 An example of module library for dilution/mixing operations [6]

Size of mixing module	Completion time (s)
2 × 4-array	2
2 × 3-array	4
2 × 2-array	6
1 × 4-linear array	3

In order to overcome the drawbacks associated with characterization, biochips integrated with sensing systems, i.e., cyberphysical microfluidic biochips, are being developed [1, 4, 9, 10]. The feedback provided by the sensing system enables real-time concentration checking, error detection, and error correction for fluidic operations [1]. Therefore, essential operations such as droplet dispensing and mixing can be precisely implemented on cyberphysical microfluidic biochips without the need for specifying a module library or characterizing a bioassay [1,9]. However, today's synthesis algorithms for mapping biochemistry protocols to the chip still rely on characterization procedures for bioassays [11–13]. Hence the advantages of cyberphysical integration are not fully exploited, and precious samples/reagents and time are consumed and wasted during characterization.

In order to map the protocol of a bioassay to the detailed implementations of fluidic operations on the biochip, bioassay synthesis algorithms need to provide all the information required for the execution of the bioassay, including the scheduling of operations, transportation paths of droplets, module placement of operations, the pin-assignment configuration of the biochip, and the wire-routing solution for the biochip. However, current synthesis algorithms cannot provide all the details listed above due to the following limitations:

1. Prior work is oblivious to variability and uncertainty in biochemical processes. The competition-time of fluidic operations in practical applications may be different from the time defined in a module library, thus the accuracy of the scheduling results for fluidic operations cannot be guaranteed.
2. On-line computation in prior work does not provide information about the paths of droplet transportation.

To overcome the above drawbacks, we propose a new design flow for cyberphysical microfluidic biochips. The key contributions and benefits of this chapter are as follows:

1. We propose the design of microfluidic biochips that uses multiple clock frequencies. The execution time of the bioassay can be reduced without additional degradation of electrodes or hardware cost.
2. In order to handle the uncertainties in completion-time of fluidic operations, we propose an "operation-interdependence-aware" synthesis algorithm—the first on-chip biochemistry synthesis procedure that does not use the module library as a design guideline. Using this algorithm, the characterization process can be removed. The algorithm results in a design approach that considers completion-time uncertainties for fluidic operations, hence it leads to the improved accuracy of fluidic operations.

3. We propose the on-line droplet-routing method that has low computational complexity, and the response time of the cyberphysical system becomes negligible. Therefore, the degeneration of intermediate products for the bioassay can be avoided.

The remainder of this chapter is organized as follows. Section 4.2 presents the design of microfluidic biochips that uses multiple clock frequencies. Section 4.3 introduces the framework of operation-dependency-aware synthesis. Based on results derived from the proposed synthesis algorithm, integrated on-line decision-making for droplet transportation path is presented in Sect. 4.4. Simulation results for three widely used bioassays are presented in Sect. 4.5. Section 4.6 concludes the chapter.

4.2 Biochips with Multiple Clock Frequencies

When an actuation voltage is applied, electrodes tend to undergo oxidation (loss of electrons), which is referred to as "electrolytic corrosion". This leads to the degradation extent of electrodes on biochips [14,15]. Experimental results published in the literature demonstrate that the degradation of an electrode is directly related to the number of times that it is switched on and off [16]. With the same sequence of electrode actuation vectors, electrodes will degrade faster under higher clock frequency. On the other hand, an increase in the clock frequency can reduce the execution time of fluidic operations [5]. Hence, in order to ensure the reliability of electrodes on the biochip, and complete the bioassay under timing constraints in the meantime, it is important to choose an appropriate clock frequency.

Fluidic operations are divided into two categories: frequency-sensitive operations and frequency-insensitive operations. The completion time of droplet transportation and dispensing is determined by clock frequency, because the droplet will be moved from one electrode to another adjacent electrode during each clock cycle; hence the rate at which a droplet is transported or dispensed is proportional to the clock frequency. Note that, if the transportation or dispensing path for a droplet consists of P electrodes, then the number of clock cycles required to dispense or move the droplet is also P. This number of clock cycles is only related to the length of the transportation path, and independent of the clock frequency. If we increase the electrode switching frequency for droplet transposition and dispensing, the time needed for these operations can be shortened. Hence, we conclude that by increasing the clock frequency, the transportation and dispensing time of droplets can be accelerated without affecting the chip reliability.

The execution times of mixing and dilution operations cannot be significantly reduced by increasing the clock frequency. For example, experimental results for droplet mixing show that, at a frequency of 8 Hz, the time spent on the mixing operation is 12 s; when the frequency is increased to 16 Hz, the mixing time decreases to 11 s [5]. Hence the mixing time only decreases 8.3 % while the rate of

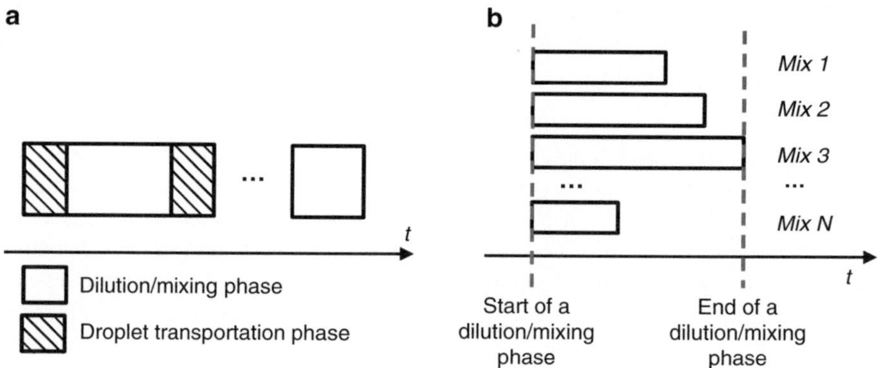

Fig. 4.1 (**a**) Droplet transportation and dilution/mixing operations are scheduled in different phases; (**b**) start and end time of a D/M phase

degradation of electrodes increases 100 %. Therefore we conclude that the lifetime of the biochip may be adversely affected by increasing the clock frequency for dilution/mixing operation, while the completion time of the operation will not be significantly reduced.

In order to minimize the completion time of a bioassay with less impact on chip reliability, and also improve the flexibility of the biochip, it is desirable to run different categories of operations at different clock frequencies. Hence we propose to schedule transportation/dispensing operations and dilution/mixing operations at different time segments. The time segment to implement droplet transportation is defined as the "transportation phase" (T phase), and the segment to implement dilution/mixing operations is defined as the "dilution/mixing phase" (D/M phase); see Fig. 4.1a. Only transportation operations or the dilution/mixing operations are carried out on the biochip for each phase.

Assume that before the execution of a bioassay we have already determined the set of dilution/mixing operations to be implemented at each D/M phase. At run-time, the biochip operates under clock frequency f_T in the T phase. Output droplets of previous steps and droplets dispensed from reservoirs are moved to the modules where the subsequent dilution/mixing operations are to be carried out. After all the droplets reach their destination modules, the biochip enters the D/M phase. The dilution/mixing operations that are scheduled in the same phase start together, and they are carried out under clock frequency $f_{D/M}$. When the feedback from sensors indicates that all the dilution/mixing operations have been completed, the D/M phase ends and the biochip enters the next T phase, as shown in Fig. 4.1b. In this way, the biochip "switches" between the T phase and the D/M phase with different clock frequencies based on feedback from sensors.

The following three methods can implement this biochip design using multiple clock frequencies with negligible extra cost, based on the hardware setup shown in Fig. 2.5.

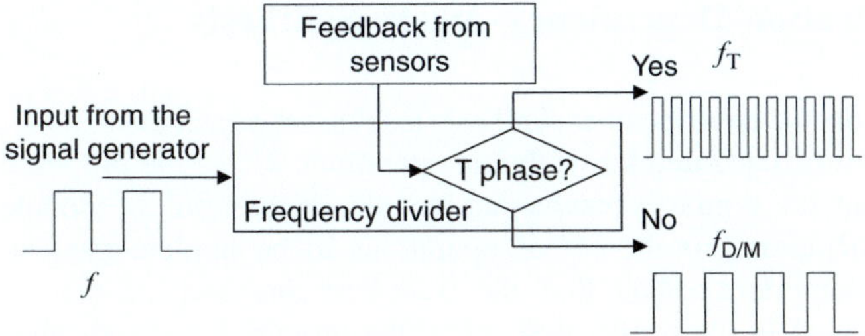

Fig. 4.2 A tunable frequency-divider controlled by the feedback from sensors

- **Method I—Software implementation:** Since the output frequency of the signal
  generator can be controlled by the software, the frequency can be adjusted
  dynamically during the execution of the bioassay. Recent work in a different
  context, viz. to understand the reliability impact of multiple frequencies, has
  demonstrated the feasibility of such a hardware setup [16].
- **Method II—Hardware implementation:** The tunable frequency-divider controlled by the feedback from sensors can generate control signals with more than
  two distinct frequencies. A tunable frequency-divider can be implemented on an
  FPGA, as shown in Fig. 4.2. It will not lead to any extra cost when handling more
  than two frequencies.
- **Method III—By adjusting the actuation signals for electrodes**: For an electrode
  that is driven by a signal generator with a fixed pulse frequency, we can adjust
  the frequency of switching on/off by changing the actuation sequences of the
  electrodes. For example, we assume that the actuation signal "1" means switching
  the electrode on, and signal "0" means switching the electrode off. Then, by
  applying a signal sequence such as "11110000", the frequency with which the
  electrode is switched on/off will be one-fourth of the frequency associated with
  the signal sequence "10101010".

In the D/M phase, the biochip operates at a nominal frequency (for example,
$f_{D/M} = 8\,\text{Hz}$), while in the T phase, higher-frequency signals (for example, $f_T =$
$16\,\text{Hz}$) are applied to the electrodes. The time spent on dispensing and transporting
droplets can be significantly reduced with no additional degradation of electrodes or
any additional cost in hardware.

It is important to note that, the above methods can be used to generate two
or more working frequencies for biochips. For simplicity, two distinct working
frequencies are used in the simulation for D/M and T phases of the bioassay in
this paper.

Since the time spent on each dilution/mixing operation is determined by sensor
feedback rather than a pre-determined module library, we can derive a "semi-
deterministic" design for biochips when considering timing uncertainties. The
design includes the synthesis result (Sect. 4.3), and droplet transportation paths in
each T phase (Sect. 4.4).

4.3 Operation-Dependency-Aware Synthesis

In this section, we describe how synthesis results can be achieved in the presence of completion-time uncertainties of fluidic operations. Using the proposed algorithm, we can obtain the synthesis results that include: (i) the result of module-placement for each operation; (ii) the set of operations to be implemented in each D/M phase. It is important to note that, the exact start time and end time of operations are not included in the results derived by the proposed synthesis algorithm; they are determined based on the feedback from sensors during the bioassay run-time. Therefore, the derived synthesis results are semi-deterministic.

The proposed synthesis algorithm focuses on the interdependency among dilution/mixing operations that are given by the sequencing graph of a bioassay. A sequencing graph is an abstract description for a bioassay; each node in the graph represents a fluidic operation, and each edge represents the interdependency for a pair of operations. For any two operations O_a and O_b, if the output droplet of O_a is the input of O_b, then there is an edge from O_a to O_b. The sequencing graph in this approach is reduced by deleting all nodes that represent dispensing operations. Two examples of sequences graphs for bioassays are shown in Fig. 4.3a, b, respectively.

A sequencing graph for a bioassay has two important properties: (i) it is a directed acyclic graph, because there is no infinite loop or repeated step in identical

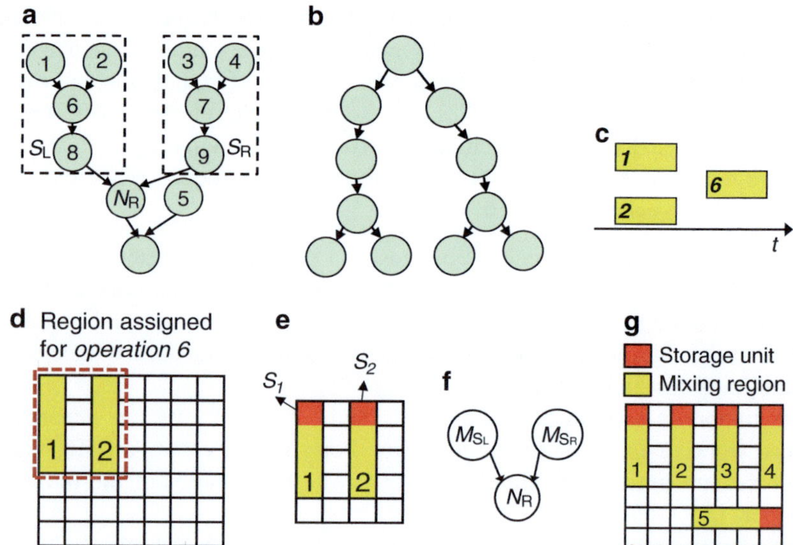

Fig. 4.3 Sequencing graph with tree structure (**a**) all edges are directed towards the root of the tree; and (**b**) all edges are directed away from the root; (**c**) scheduling results for mixing operations *1, 2,* and *6*; (**d**) module-placement for *1, 2* and *6*; (**e**) storage units S_1 and S_2 inside the modules assigned for operations *1* and *2*; (**f**) packaged macro-operations M_{S_L} and M_{S_R}; (**g**) a feasible solution for module-placements of operation *1* $\sim$ *5* on an 8 × 8 array

conditions for an assay protocol; (ii) the numbers of input droplets and output droplets for each operation are no more than 2. Therefore, the in-degree and out-degree of each node in the sequencing graph are at most 2.

In the following parts, we first introduce the synthesis algorithm for sequencing graphs with the structure of directed-trees, and then introduce the steps for synthesizing a bioassay in the general case.

4.3.1 Synthesis for Sequencing Graphs with Directed Tree Structure

First we study bioassays whose sequencing graphs are directed trees. Two examples can be found in Fig. 4.3a, b. Note that the edges in a directed tree can be directed either towards or away from a particular vertex. The node without parent nodes in Fig. 4.3a and the node without children nodes in Fig. 4.3b are defined as the leaf nodes of the directed trees. The operations represented by these leaf nodes are defined as "the lowest-level operations" of the bioassay. The node without children nodes in Fig. 4.3a and the node without parent nodes in Fig. 4.3b are defined as the root nodes of the directed trees.

The difference between these two sequencing graphs is that, all edges in Fig. 4.3a are directed towards the root of the tree, while in Fig. 4.3b, all edges are directed away from the root. After switching the directions of all of the edges, the directed tree in Fig. 4.3b can be analyzed in the same way as the one in Fig. 4.3a. Thus, we use the structure with all edges directed towards the tree root (shown in Fig. 4.3a) to analyze the proposed synthesis procedure.

In the synthesis procedure, starting from the lowest-level operations, the schedule and module-placement are determined for the operations on a level-by-level basis. Consider the following example. Operations 1, 2, and 6 are three mixing operations shown in Fig. 4.3a. Here 1 and 2 are the lowest-level operations and their outputs are the inputs of operation 6, hence 1 and 2 must be completed before 6 starts. Based on the interdependency relationship, operations 1 and 2 are implemented in the same D/M phase, while the operation 6 is scheduled to be implemented in the next D/M phase, as shown in Fig. 4.3c. If we write the set of operations to be implemented at the ith D/M phase as S_i, then the schedules of operations 1, 2, and 6 can be written as: $\{1, 2\} \subseteq S_1$ and $\{6\} \subseteq S_2$.

Next, mixing operations 1 and 2 are mapped to two mixers on the biochip. The sizes of widely-used mixers on digital microfluidic biochip can be found in Table 4.1 [6]. As in prior work [6, 11, 13], segregation cells are set as wrappers for each mixer in order to isolate droplets that are being manipulated concurrently on the biochip. Since the mixing operation 6's inputs are the outputs of operations 1 and 2, we can refer to the region that overlaps with the modules for 1 and 2 as the "execution region" of operation 6, as shown in Fig. 4.3d. After the mixing operation has been completed for each mixer, the product droplet stays inside the mixer until the end of the D/M phase, i.e., part of the mixer works as a "storage" unit. Therefore, in the synthesis approach, there is no need to assign specific storage modules. An example is as follows. S_1 and S_2 in Fig. 4.3e represent the storage units inside the mixers assigned to operations 1 and 2, respectively. Since the execution region of operation 6 overlaps with S_1 and S_2, the outputs of mixing operations 1 and 2 can be directly "fed" into operation 6 with no module-to-module transportation. After operation 6 is completed, the storage unit for the product droplet of operation 6 will overlap with either S_1 or S_2.

In this way, the module-placement and schedule for operations 1, 2, and 6 are obtained. After operation 6 is finished, the execution region of operation 6 will be assigned for operation 8.

Next, operations 1, 2, 6, and 8 are packaged as a "macro-operation" M_{S_L}; similarly, operations 3, 4, 7, and 9 are packaged as M_{S_R} as shown in Fig. 4.3f. The schedule for operations $1 \sim 4$ and $6 \sim 9$ are: $\{1, 2, 3, 4\} \subseteq S_1$, $\{6, 7\} \subseteq S_2$, and $\{8, 9\} \subseteq S_3$.

Similarly, the placement of modules for M_{S_L}, M_{S_R} and N_R in Fig. 4.3f can be determined based on their interdependency. The module for M_{S_L} will be placed "beside" the region occupied by M_{S_R}, and after the completion of M_{S_L} and M_{S_R}, N_R will be mapped to the same region that is assigned to M_{S_L} and M_{S_R}. In this way the schedule and resource assignment results for all the operations shown in Fig. 4.3a can be derived level by level.

Suppose we arbitrarily pick a node N_R from the directed-tree structure shown in Fig. 4.3a, and label the set of operations on the left and right sub-trees of N_R as S_L and S_R, respectively. Then the derived synthesis results of the proposed operation-interdependency-aware synthesis algorithm have the following three characteristics:

Characteristic 1: Operations in S_L and S_R are executed in two separate regions of the biochip, i.e., they do not share any on-chip resources.
Characteristic 2: Assume that E_{S_L} and E_{S_R} are the sets of electrodes where operations in S_L and S_R are conducted, respectively, and E_{N_R} is the set of electrodes that is assigned to operation N_R, then we have: $E_{N_R} \subseteq (E_{S_L} \cup E_{S_R})$.
Characteristic 3: If the storage units assigned for operations in S_L and S_R are written as S'_{S_L} and S'_{S_R}, respectively, and the storage unit assigned to operation N_R is written as S'_{N_R}, then we have: $S'_{N_R} \subseteq (S'_{S_L} \cup S'_{S_R})$.

Based on Characteristic 2, we conclude that the resource bound to an operation (i.e., the module-placement) is determined in turn by its predecessor operations. For a sequencing graph with a directed-tree structure, the resources for other operations can be easily determined when resources assigned to operations represented by leaf nodes are established.

For instance, the synthesis result for the sequencing graph shown in Fig. 4.3a can be determined as follows. According to the proposed resource binding steps, the modules for operations 1 and 2 will be placed beside each other, and the modules for operations 3 and 4 also will be placed beside each other. Since operations that are scheduled in the same D/M phase are concurrently performed, we assume that the size of the mixers allocated to all these operations are also the same. Without

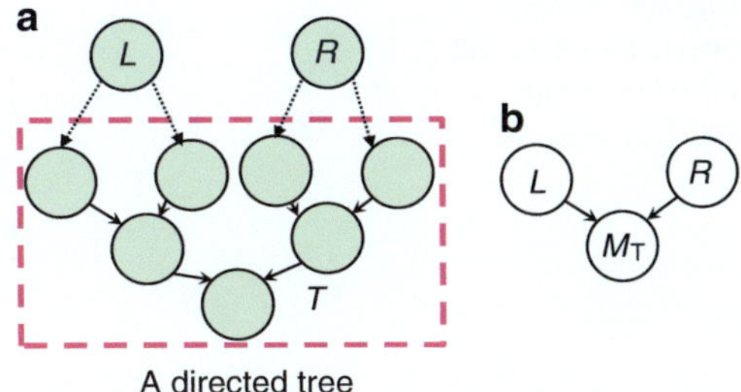

Fig. 4.4 (**a**) Partitioning of a sequencing graph; (**b**) the interdependency of macro-operation M_T, operation L and operation R

loss of generality, we assume that operations $1 \sim 5$ are executed in 1×4 mixers. The corresponding result for the placement of the modules on a 7×7 array is shown in Fig. 4.3g. Each mixing module has one storage unit inside. When the mixing operation is completed, the product droplet stays inside the corresponding storage unit. Based on the module-placement for operations $1 \sim 5$, module-placement for other succeeding operations can be obtained. For example, the regions where operations 1 and 2 are implemented will be assigned to their successor operations. In this way, the module placement of operations in the directed-tree structure are determined. The schedules of operations in the directed-tree can also be determined based on their input/output interdependency.

4.3.2 Synthesis for Sequencing Graphs in General Cases

For sequencing graphs that are not directed trees (i.e., there are nodes whose out-degrees are greater than 1, as shown in Fig. 4.4a), we can derive their synthesis results using the following steps:

1. **Graph partitioning:** First we determine the set of nodes $\{N_{n_1}, N_{n_2}, \ldots, N_{n_k}\}$ whose out-degrees are greater than 1, then remove all the edges directed away from these nodes. Then the graph is partitioned into multiple directed trees $\{T_1, T_2, \ldots, T_n\}$, and each node in $\{N_{n_1}, N_{n_2}, \ldots, N_{n_k}\}$ becomes the root node in a directed tree.
2. **Synthesis for directed trees:** We apply operation-interdependency-aware synthesis to each directed tree, and derive the corresponding synthesis results.
3. **Sorting of directed trees:** For any pair of trees T_A and T_B, we suppose there exists a node $O_{T_{A_1}} \in T_A$ and a node $O_{T_{B_1}} \in T_B$, such that $O_{T_{A_1}}$ is the predecessor of $O_{T_{B_1}}$. Then the relationship between trees T_A and T_B is expressed as $T_A < T_B$. Any two elements T_x and T_y in $\{T_1, T_2, \ldots, T_n\}$ may stand in any of three mutually exclusive relationships to each other: $T_x < T_y$, or $T_x > T_y$, or $T_x = T_y$ (neither of the other two). The conflicting case (i.e., "$T_x < T_y$ and $T_x > T_y$") will never occur. The proof can be found in Lemma 4.1 of this section.
4. **Merge the synthesis results:** The synthesis result of $\{T_1, T_2, \ldots, T_n\}$ are merged together according to their order. If $T_x < T_y$ then operations in T_x

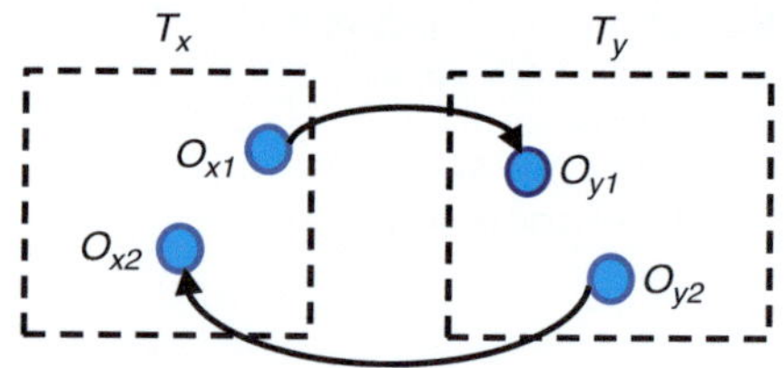

Fig. 4.5 Assume two directed trees T_x and T_y have the relationships that $T_x < T_y$ and $T_x > T_y$

must be implemented before T_y; if $T_x > T_y$ then operations in T_x must be implemented after T_y; if $T_x = T_y$ then operations in T_x and T_y are implemented in parallel.

Based on the method of graph partitioning, we have the following lemma:

Lemma 4.1. *For any two elements T_x and T_y in the set of directed tree $\{T_1, T_2, \ldots T_n\}$ partitioned from a sequencing graph, the conflict relationship "$T_x < T_y$ and $T_x > T_y$" never occurs.*

Proof. This lemma can be proved by *reductio ad absurdum*. First we assume that these two directed tree have the relationships $T_x < T_y$ and $T_x > T_y$. Then as shown in Fig. 4.5, we notice that two nodes $O_{x_1} \in T_x$ and $O_{y_1} \in T_y$, such that O_{x_1} is the predecessor of O_{y_1}; we can also find two node $O_{x_2} \in T_x$ and $O_{y_2} \in T_y$, such that O_{y_2} is the predecessor of O_{x_2}. Without loss of generality, here we draw the node O_{x_1} as the immediate predecessor of O_{y_1}, and the node O_{y_2} as the immediate predecessor of O_{x_2}.

Next we can prove that O_{x_1} is the root node for T_x. As introduced in Sect. 4.4, we first find out all the nodes $\{N_{n_1}, N_{n_2}, \ldots N_{n_k}\}$ whose out-degrees are more than 1. Then all the edges that are directed away from these nodes are removed. Hence in the trees we derived from the partitioning procedure, the out-degrees of all the nodes except the root node are equal to 1. The out-degree of the root node is equal to 0. Assume O_{x_1} is not the root of T_x. Then in T_x, there must be an edge from O_{x_1} to another node in T_x. As shown in Fig. 4.5, there is another edge from node O_{x_1} to O_{y_1}. Therefore, the out-degree of O_{x_1} in the original sequencing graph (i.e., the sequencing graph before being partitioned) is equal to 2. □

Based on the step of graph partitioning discussed above, O_{x_1} whose out-degree is equal to 2, must be a root node for a directed tree derived. Hence we have reached the conclusion that O_{x_1} is the root node for T_x. Similarly, we can reach the conclusion that O_{y_2} is the root node for T_y.

Then based on the characteristic of directed tree, there must exist a directed path from the node O_{x_2} toward the root node O_{x_1} in T_x, and there must exist a directed path from the node O_{y_1} towards the root node O_{y_2} in T_y. Since there are the edge from O_{x_1} to O_{y_1}, and the edge from O_{y_2} to O_{x_2}, we get a directed circle that connects the nodes O_{x_1}, O_{y_1}, O_{y_2}, and O_{x_2} in the sequencing graph. However, as introduced in Sect. 4.3, a sequencing graph for bioassay do not have a directed cycle.

1: Partition a sequencing graph G into directed-trees $\{T_1, T_2, ..., T_n\}$;
2: for each $T_i \in \{T_1, T_2, ..., T_n\}$ do
3:    Start from leaf nodes, determine relative positions of dilution/mixing modules on a level-by-level basis;
4:    Determine schedules of operations;
5:    Package operations the entire directed tree as a "macro-operation" M_{T_i};
6: end for
7: Sort the directed trees based on operation interdependencies;
8: Merge synthesis results of directed trees based on their orders.

Fig. 4.6 Pseudocode for operation-interdependency-aware synthesis

Hence we have proved that, the conflict relationships between T_x and T_y will not exist. This completes the proof of the lemma.

From the above steps, the synthesis results can be derived for general sequencing graphs. An example is shown in Fig. 4.4a. The graph is divided into three parts: the node that represents operation L, the node that represents operation R, and a directed tree T. The operations in T are packaged as a "macro-operation" M_T. As operation L and R are both predecessors of operations in T, the interdependency of macro-operation M_T, operation L and operation R is show in Fig. 4.4b. According to the above discussion, the relationship among R, L, and M_T can be written as $R = L > M_T$.

Therefore, operations L and R will be concurrently executed, and operations in M_T will be executed after operations L and R are both completed. The corresponding module-placement result can be derived in a similar way as the sequencing graph shown in Fig. 4.3f.

The operation-interdependency-aware algorithm described above can derive the semi-deterministic synthesis result for a bioassay. The synthesis result assigns each dilution/mixing operation to a specific D/M phase, while completion-times of these operations are determined on-line during the execution of the bioassay. The robust module-placement thus derived is independent of the execution time for each operation. When timing uncertainties of dilution/mixing operations exist, the module-placement of a bioassay remains unchanged. The pseudocode for the entire operation-interdependency-aware synthesis approach is shown in Fig. 4.6. By applying the proposed synthesis approach, the completion-time uncertainties are handled in a reliable manner by on-line decision-making during the execution of the bioassay, even though the completion-time of fluidic operations are unpredictable before the bioassay is run on the fabricated biochip.

4.4 Droplet-Routing Procedure

The practical application of cyberphysical microfluidics requires control software to ensure the synthesis results of the bioassay are routable, and also, to determine droplet transportation paths for each droplet transportation phase.

As introduced in Sect. 4.2, in the D/M phase, the biochip will stop an operation when the feedback from sensors indicates the operation has been finished. Then the output droplet will be stalled inside the modules, and will wait to be used as the input of the following operation in the next D/M phase. In each T phase, all the dilution/mixing operations executed in the last D/M phase have already finished, and their output droplets of operations are stalled in their modules. The electrodes where these droplets stay, as well as their neighboring electrodes, are considered as "obstacles" for droplet transportation. The droplet transportation paths cannot overlap with the obstacles, otherwise the interference between droplets may occur.

The transportation of droplets during a bioassay consists of three parts: moving droplets from one module to another module; moving droplets from on-chip reservoirs to the modules; and moving the "extra product droplets" to waste reservoirs. For some bioassays, it is important to note that not all of the intermediate product droplets are used for subsequent operations. Those product droplets that are not used in the bioassay are defined as extra product droplets, and must be collected in the waste reservoir.

As introduced in Sect. 4.3, the placements of modules for dilution/mixing operations are determined based on their input/output interdependencies. For any droplet that contains an intermediate product of the bioassay, the module where the droplet is generated and the module where it will participate during the next operation, are overlapped. The output droplets for each module only need to "wait" at the original position; they will be directly "fed" into the module of the following operation. Therefore, the transportation of droplets from modules to modules can be eliminated. Only droplet transportations from reservoirs to modules and from modules to the waste reservoir need to be considered in the T phase.

In this section, first we consider the routability for the result derived by the operation-interdependency-aware synthesis approach. Then the method of online decision-making on the transportation paths of droplets is introduced.

4.4.1 Routability Analysis

If we randomly select an operation O_N from the sequencing graph, and the immediate predecessor operations of O_N are written as O_L and O_R, then we have the following lemma:

Lemma 4.2. *Transportation paths from a reservoir R to the module of O_N exist if there are transportation paths from the reservoir to the modules of immediate predecessor operations O_L and O_R.*

Proof. The existence of transportation paths from the reservoir R to the modules of operations O_L and O_R indicates that paths from the reservoir to the modules can be found when there are obstacles R_{LR} on the biochip. Here, the set of storage units and their neighboring electrodes is written as R_{LR}. Even if no droplet stays on the storage unit, we still consider the "empty" storage unit (and its neighboring electrodes) as obstacles.

The set of obstacles is written as R_N when we search for the transportation path from the reservoir to the module of operation O_N. Based on Characteristic 3 of the results derived by the operation-interdependency-aware synthesis procedure proposed in Sect. 4.3, the droplet that stays in module O_N is restricted inside one of the regions that have ever been assigned as storage units for operation O_L and O_R. $\quad\square$

Thus we have $R_N \subseteq R_{LR}$, i.e., each electrode in the obstacle R_N is contained by obstacle R_{LR}.

We have already assumed that the routing problem has a solution with the existence of obstacle R_{LR}. Therefore, with the existence of R_N, there must exist a routing solution from the reservoir to the module of O_N. This completes the proof of the lemma.

During the execution of bioassays, the input droplets for O_N can be from different reservoirs, while it is important to note that Lemma 4.2 is insufficient to ensure the existence of the transportation from any on-chip reservoir to the module for O_N. Therefore, in order to guarantee that droplets dispensed from any on-chip reservoir can be transported to the module of O_N, the droplet-routing paths between reservoirs need to be considered.

In the synthesis result of a bioassay, for each operation, if there exist transportation paths to move its input droplets into the module and there exist transportation paths to move the extra droplets into the waste reservoir, then the synthesis result is "routable". By applying Lemma 4.2 to the sequencing graph with a directed-tree structure, as well as considering the differences between reservoirs, we can further derive the following lemma that gives the sufficient conditions for routability of synthesis results:

Lemma 4.3. *For a given bioassay whose sequencing graph is a directed-tree, its corresponding synthesis result is routable if the following two constraints are satisfied:*

Constraint 1. There exist transportation paths that connect all the modules of the lowest-level operations and their corresponding reservoirs. Here the "lowest-level operations" is defined as "the operations that are represented by leaf nodes in the sequencing graph of bioassay".

Constraint 2. There exist transportation paths that connect all the on-chip reservoirs.

Proof. We write the set of modules corresponding to the lowest-level operations in the sequencing graph (M_{leaf}) as $\{M_{L_1}, M_{L_2}, \ldots, M_{L_N}\}$, and the set of reservoirs R^* as $\{R_1, R_2, \ldots, R_r\}$.

For each reservoir R_i, the sequencing graph of the bioassay defines the set of modules whose input droplets are from R_i. For example, if the input sample/reagent of $\{M_{L_{i_1}}, M_{L_{i_2}}, \ldots, M_{L_{i_k}}\}$ is stored in reservoir R_i, then the droplets dispensed from R_i need to be transported to these modules.

Constraint 1 ensures that there exist transportation paths from reservoir R_i to modules $\{M_{L_{i_1}}, M_{L_{i_2}}, \ldots, M_{L_{i_k}}\}$. Constraint 2 ensures that the reservoirs $\{R_1, R_2, \ldots, R_r\}$ are interconnected by transportation paths. Hence for any arbitrarily chosen reservoir in R^*, and any arbitrarily chosen module from M_{leaf}, there exist transportation paths between them.

Then based on Lemma 4.1, we know that there exist transportation paths that can feed droplets from any reservoir in R^* to any immediate successor operation of these lowest-level operations.

By applying Lemma 4.1 level-by-level to the sequencing graph until we reach the root node, we find that there exist transportation paths from a reservoir in R^* to the module corresponding to any operation in the bioassay. Hence for each operation, there exist transportation paths from the corresponding reservoirs to the module of the operation. Therefore, the synthesis result is routable. This completes the proof of the lemma.   □

4.4.2 Searching Droplet-Routing Paths

As discussed in Sect. 4.3, the synthesis results derived by the operation-interdependency-aware synthesis procedure is independent of the execution time of operations. For each T phase, the corresponding module-placement configuration remains unchanged when the execution time of the operations varies. The routability of the synthesis result, the starting and ending positions of each droplet, as well as the positions for storage unites of droplets, have already been determined before the execution of a bioassay. With this available information, the droplet transportation paths for each T phase can be derived by the algorithms proposed in previous publications [17–20].

4.4.3 Online Decision-Making for Droplet-Routing

By applying the operation-interdependency procedure discussed in Sect. 4.3 and the computation steps for droplet transportation as discussed above, we can derive the following information before the execution of a bioassay: (i) schedule of operations, i.e., the set of operations to be implemented in each D/M phase; (ii) module-placement for operation that is not affected by timing uncertainties in the execution of fluidic operations; (iii) droplet transportation paths for each T phase.

Since the uncertainties about execution times for the fluidic operations have no influence on the routability of the synthesis results or the transportation paths of droplets, routing paths can be derived before the execution of the bioassay by applying A^* algorithm [17]. The response time of on-line decide-making for the cyberphysical system is not affected by the computational complexity of the $\{A^*$ droplet routing algorithm.

Based on the pre-determined module-placement results and droplet transportation paths, the bioassay can be performed under the completion-time uncertainties and yet provide high accuracy in the final reaction outcome.

4.5 Simulation Results

In this section, we first compare the proposed uncertainly-aware synthesis algorithm with previous work. Then the simulation results derived under different clock frequencies are presented.

4.5.1 Comparisons Between the Proposed Synthesis Algorithm and Previous Algorithms

In this part, we evaluate and compare our results with prior work on biochip synthesis [6,12,13] and a recently published cyberphysical software-based recovery method based on a greedy algorithm [11, 21]. All these baseline methods are oblivious of timing uncertainties in fluidic operations. The proposed synthesis algorithm is compared with prior work in the following aspects: (i) the number of droplets consumed; (ii) the yield of bioassay; (iii) the response time in the presence of timing uncertainties; and (iv) the number of operations that are interrupted from the completion-time uncertainties.

4.5.2 Number of Droplets Consumed

For the synthesis algorithm proposed in [21,22], bioassays are characterized before they are executed on the biochip. In the characterization procedure, each operation needs to be executed at least three times [9]. The comparison of the number of droplets consumed for each bioassay in [12, 21, 22] and the proposed design is shown in Fig. 4.7. Note that the prior methods use the same number of droplets each. We find that the number of droplets is significantly reduced in the proposed design.

4.5.2.1 Yield Estimation for Biochips with No Feedback-Based Adaptation

As discussed in Sect. 4.1, the execution time of a fluidic operations needs to be considered as a random variable rather than a known constant. If no feedback-based control is used and the synthesis of the biochip only relies on a module library with

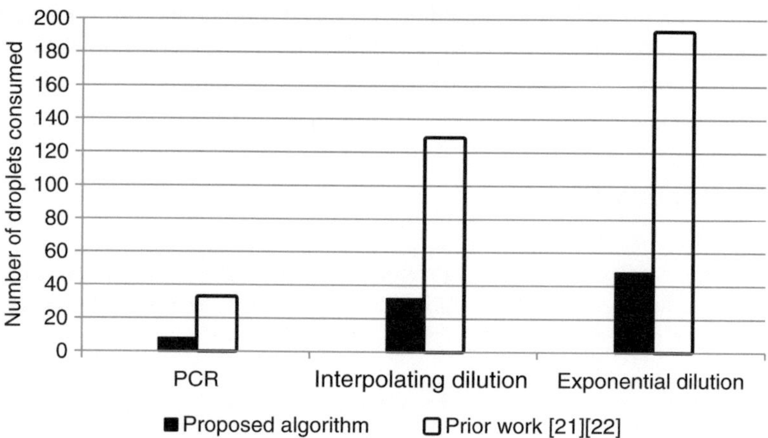

Fig. 4.7 Comparison between the number of droplets consumed in the biochips of [21, 22] (droplet consumptions are the same for the methods of these two papers) and the proposed method

constant operation times, there is a high likelihood that many operations cannot finish within the allocated time. The outputs of these "unfinished operations" will be unqualified droplets, and the outcome of the bioassay will be unacceptable.

Next we compute the probability that a bioassay fails due to the occurrence of unfinished operations. Assume that the execution time for each operation is Gaussian random variable, and for any two operations in a bioassay, their execution times are independent of each other. Suppose that for any operation O_i, its completion time T_i has mean value μ_i and variance σ_i^2. If we define the time spent on operation O_i as $T_i = \mu_i + \sigma_i$, and the real execution time T_i^* does not exceed T_i, then O_i will be correctly executed; otherwise O_i is deemed to have failed. A bioassay terminates with an acceptable outcome only if all the operations are executed correctly. Let P_i be the probability that a dilution/mixing operation O_i is executed correctly, and let the number of dilution/mixing operation in a bioassay be N_b. Then the probability P_{success} that the bioassay is successfully implemented can be expressed as: $P_{\text{success}} = \prod_{i=1}^{N_b} P_i = P^{N_b}$ (if $P_i = P$ for any i).

Therefore, the probability of successful implementation P_{success} exponentially decreases with the increasing number of dilution/mixing operation. Hence for a realistic bioassay with a large number of fluidic operations, the compound yield will be unacceptably low.

Next, based on above discussion, we compute the numerical values of P_{success} for three widely used laboratory protocols, namely exponential dilution of a protein sample, interpolation dilution of a protein sample, and polymerase chain reaction (PCR). The sequencing graph and the detailed description of the protocol for these bioassays can be found in [6].

Let T_i be the time spent on operation O_i when a static synthesis method is applied based only on a module library, and P_i be the probability that O_i is successfully

Table 4.2 Probabilities of bioassays being successfully implemented without unfinished operations

Bioassay	No. of dilution/ mixing operations	P_{success}, T_i listed		
		μ_i	$\mu_i + \sigma_i$	$\mu_i + 2\sigma_i$
PCR	7	0.01	0.30	0.85
Interpolating dilution	35	~ 0	~ 0	0.45
Exponential dilution	47	~ 0	~ 0	0.34

completed before the next operation starts. Based on the characteristics of the Gaussian distribution, if we set the time spent on operation O_i as $T_i = \mu_i + \sigma_i$, then $P_i = 0.84$; if we set $T_i = \mu_i + 2\sigma_i$, then $P = 0.98$. Table 4.2 lists the probabilities that bioassays are successfully completed (i.e., with no unfinished operations) when no cyberphysical adaptation is used. When we conservatively set $T_i = \mu_i + 2\sigma_i$ (and increase the operation times considerably) and obtain $P_i = 0.98$, it leads to low probability (0.34) of implementing the exponential mixing of protein bioassay successfully. We can further calculate that, in order to improve the yield of the exponential dilution bioassay to 0.90 and above, we need to set $T_i = \mu_i + 4\sigma_i$.

We therefore conclude that the bioassay execution on conventional biochip platforms without a sensing system, closed-loop control, or uncertainly-aware synthesis, will lead to unacceptably low bioassay yield and low confidence in reaction outcomes.

From Table 4.2, we also find that P_{success} can be increased by increasing T_i. However, on the other hand, increasing T_i will elongate the reaction completion time, and increase the risks of excessive heating and evaporation of droplets [23]. Therefore, T_i should be set within in a reasonable range. Due to the lack of sufficient data thus far from real experiments on fabricated chips, the additional probability of bioassay failure caused by excessive heating and evaporation is not considered here. In this way, we do not quantify this important shortcoming of the baseline methods that we use for comparison.

Next we compare the bioassay completion time of the proposed operation-interdependency-aware synthesis algorithm with the parallel recombinative simulated annealing (PRSA)-based synthesis algorithm [6, 24]. Here the module library shown in Table 4.1 is used [6], and simulations are executed by considering timing uncertainty. We assume that for each operation, its average execution time μ_i is the time defined in the library, and σ_i is $0.1\mu_i$. For example, for a dilution operation executed on a 2×3 array, $\mu_i = 6$ s, and $\sigma_i = 0.6$ s.

In order to increase the yield for the baseline design, when running the PRSA-based algorithm, extra execution time ΔT_i is added for each operation. In the simulations, we assume that the PCR bioassay, the exponential dilution bioassay, and the interpolation dilution bioassay are all executed on an 8×8 direct-addressing array.

The simulation results with ΔT_i, which are set to 0, σ_i, $2\sigma_i$, and $4\sigma_i$, are shown in Table 4.3. The completion time of the bioassays derived by PRSA-based algorithm

Table 4.3 Bioassay completion times derived by PRSA-based algorithm [6]

Bioassay	Completion time (s) derived by PRSA-based algorithm [6] ΔT_i listed			
	0	σ_i	$2\sigma_i$	$4\sigma_i$
PCR	26	28	31	35
Interpolation dilution	177	184	195	201
Exponential dilution	195	196	202	208

increases with ΔT_i. It is important to note that, the synthesis result derived by the PRSA-based algorithm is deterministic, i.e., the synthesis results determine the start and stop times of fluidic operations before running bioassays on the fabricated biochip.

In contrast, the synthesis result derived by the proposed method is semi-deterministic, the exact start and stop times of fluidic operations will be determined by feedback from the sensing system when running the bioassays on the fabricated biochip. If we assume the completion-times of fluidic operations are exactly the same with the module library defined in [6], then the completion times for PCR bioassays, interpolation dilution bioassays, and exponential dilution bioassays will be 25, 158, and 182 s, respectively.

4.5.2.2   Response Time in the Presence of Timing Uncertainties

In the re-synthesis algorithm proposed in [11], when a sensor detects that the status of a droplet is inconsistent with the expected result, e.g., an operation is finished within the pre-determined time, a resynthesis procedure will be performed. The control software dynamically updates the synthesis steps using a greedy algorithm, and the operation with unexpected completion time will be re-executed in the resynthesis result. To compare with [11], we first randomly select a fluidic operation, and assume its execution time is longer than the time defined in the module library. Then after the resynthesis procedure is triggered, the response time for the resynthesis procedure in [11] is recorded. Here the response time is defined as the CPU time spent in deriving resynthesis solutions when the timing overshoot is detected. During on-line resynthesis, all fluidic operations for the bioassay are suspended.

The simulation is repeated 20 times (using different operations with uncertainty injected each time). The results show that the proposed algorithm has almost zero response time while the greedy approach in [11] requires around 1 s.

Table 4.4 Comparisons for the numbers of interrupted operations between PRSA-based algorithm [6] and the proposed method

Bioassay	Total no. of operations	No. of operations interrupted [6]	No. of operations interrupted (proposed method)
Exponential dilution	103	97	0
Interpolating dilution	71	65	0
PCR	15	3	0

4.5.2.3 Number of Operations Interrupted Under Uncertainty

Suppose we run the synthesis results derived by the PRSA-based algorithm on a cyberphysical biochip. If the execution time of an operation O_{longer} is longer than the time defined by the module library, then the controller can stop all other operations until operation O_{longer} is completed. In this method, time-consuming online resynthesis can be avoided. However, the executions of other operations have to be interrupted.

Based on the Gaussian assumption in Sect. 4.5.2.1, the probability that "an operation's execution time is longer than the time defined in the module library" is 0.5, if no extra execution time is added for each operation. We simulate the bioassay and count the number of operations that must be interrupted. During the execution of a bioassay, an operation may be interrupted multiple times; however, we count such cases as one interrupted operation. Hence the number of interruptions reported for the baseline method is less than the actual number of interruptions experienced.

The total number of operations, and the number of operations that are interrupted in the three bioassays, are listed in Table 4.4. We note that nearly all the operations are interrupted for the synthesis results derived using the PRSA-based algorithm. The interruptions that occur during the execution of the bioassay may influence the quality of output droplets [11]. On the other hand, in the proposed algorithm, no operations are interrupted, and the assay proceeds in an unimpeded manner.

4.5.3 Results Derived by the Operation-Interdependency-Aware Synthesis Approach

In this part, we assume that the completion times of mixing operations performed by different kinds of mixers are the same as the time defined in the module library in Table 4.1 [6]. If the bioassays are mapped to direct-addressing biochips, the synthesis results derived by the operation-interdependency-aware synthesis approach can be found in Table 4.5. Here the completion time of D/M phases is defined as the sum of time spans for all the D/M phases in the bioassay. The completion time of T phases is defined as the sum of time spans for all the T phases in the bioassay. The clock frequency of T phase is set as 1 Hz. From Table 4.5, we

Table 4.5 Synthesis results derived by the operation-interdependency-aware synthesis approach on direct-addressing biochips

Bioassay	Size of the biochip	Completion time (s)		
		D/M phases	T phases	Total
	8×8	57	125	182
Exponential	9×9	32	106	138
dilution	10×8	37	117	154
bioassay	12×8	22	145	167
	8×8	44	114	158
Interpolation	9×9	29	102	131
dilution	10×8	32	114	146
bioassay	12×8	21	130	151

can find that when the size of the biochip grows from 8×8 to 12×8, the time spent on droplet dilution/mixing decreases while the time spent on droplet transportation increases.

4.5.4 Completion Time with Multiple Clock Frequencies

For the biochip with multiple clock frequencies, if we increase the clock frequency for T phase, then the execution time for the bioassay is reduced. When bioassays are executed on an 8×8 direct-addressing biochip, the relationship between execution time of bioassays and the clock frequency of the T phase is shown in Fig. 4.8.

By increasing f_T from 1 to 10 Hz, the execution time for exponential dilution bioassay is shortened from 182 to 70 s. The completion time for interpolation dilution and PCR are also listed in Fig. 4.8.

From the simulation results, we find that when the clock frequency of the T phase increases, the completion times of PCR bioassay does not decrease significantly. This is because the time for droplet transportation in the PCR bioassay is relatively low.

4.6 Chapter Summary and Conclusions

In this chapter, we have shown how cyberphysical integration in digital microfluidics can be used to carry out on-chip bioassays despite the timing uncertainties inherent in fluidic operations such as mixing, dilution, and thermal cycling. We have presented an operation-interdependency-aware synthesis method that is responsive to such uncertainties. The proposed design approach facilitates dynamic on-line

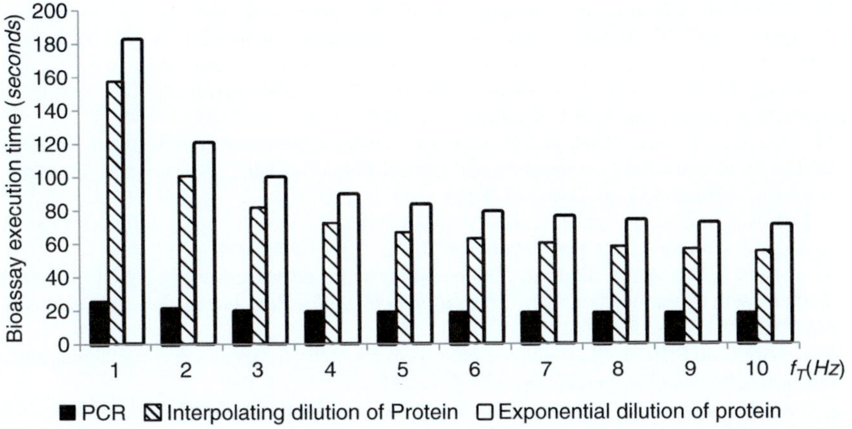

Fig. 4.8 The relationship between completion time of bioassays and f_T on an 8×8 direct-addressing biochip

decision-making for the execution of fluidic operations and droplet-routing in response to detector feedback. We have also incorporated the use of multiple clock frequencies to accelerate the time-to-response without any adverse impact on electrode reliability. We have used three common laboratorial protocols to demonstrate that, compared to uncertainty-oblivious biochip synthesis, the proposed dynamic synthesis approach decreases the likelihood of erroneous reaction outcomes, and it leads to reduced time-to-result, less repetition of reaction steps, and less wastage of precious samples and reagents. This work is therefore a step forward towards fully automated on-chip biochemistry with reliable assay outcomes and low cost.

References

1. B. Hadwen, G. Broder, D. Morganti, A. Jacobs, C. Brown, J. Hector, Y. Kubota, and H. Morgan, "Programmable large area digital microfluidic array with integrated droplet sensing for bioassays", *Lab on a Chip*, pp. 3305–3313, 2012.
2. Z. Hua, J. Rouse, A. Eckhardt, V. Srinivasan, V. Pamula, W. Schell, J. Benton, T. Mitchell, and M. Pollack, "Mutiplexed real-time polymerase chain reaction on a digital microfluidic platform", *Anal. Chem.*, vol. 82, pp. 2310–2316, 2010.
3. S. Park, P. Wijethunga, H. Moonb, and B. Han, "On-chip characterization of cryoprotective agent mixtures using an EWOD-based digital microfluidic device", *Lab on a Chip*, pp. 2212–2221, 2011.
4. M. Schertzer, *Characterization of the Motion and Mixing of Droplets in Electrowetting on Dielectric Devices*, PhD thesis, University of Toronto, Toronto, CA, 2010.
5. P. Paik, V. Pamula, and R. Fair, "Rapid droplet mixers for digital microfluidic systems", *Lab on a Chip*, vol. 3, pp. 253–259, 2003.
6. K. Chakrabarty and F. Su, *Digital Microfluidic Biochips: Synthesis, Testing, and Reconfiguration Techniques*, Boca Raton, FL: CRC Press, 2006.

7. O. Levenspiel, "Chemical reaction engineering", *New York: Wiley*, 1999.
8. M. Iyengar and M. McGuire, "Imprecise and qualitative probability in systems biology", *International Conference on Systems Biology*, Long Beach, California, 2007.
9. J. Gong, *Portable Digital Microfluidic System: Direct Referencing EWOD Devices and Operating Control Board*, PhD thesis, UCLA, 2007.
10. M. Schertzer, R. Ben-Mrad, and P. Sullivan, "Using capacitance measurements in EWOD devices to identify fluid composition and control droplet mixing", *Sensors and Actuators B: Chemical*, volume 145, pp. 340–347, 2010.
11. Y. Luo, K. Chakrabarty, and T.-Y. Ho, "A cyberphysical synthesis approach for error recovery in digital microfluidic biochips", *Proc. DATE*, pp. 1239–1244, 2012.
12. M. Alistar, P. Pop, and J. Madsen, "Online synthesis for error recovery in digital microfluidic biochips with operation variability", *Symposium on Design, Test, Integration and Packaging of MEMS/MOEMS*, pp. 53–58, 2012.
13. D. Grissom and P. Brisk, "Fast online synthesis of generally programmable digital microfluidic biochips", *Proc. CODES+ISSS*, pp. 413–422, 2012.
14. J. Gong and C.-J. Kim, "Direct-referencing two-dimensional-array digital microfluidics using multi-layer printed circuit board", *J Microelectromech Syst.*, vol. 17, Issue 2, pp. 257Ű-264, 2008.
15. Q. Al-Gayem, H. Liu, A. Richardson, N. Burd, and M. Kumar, "An on-line monitoring technique for electrode degradation in bio-fluidic microsystems", *IEEE International Test Conference*, pp. 654–663, 2010.
16. L. Huang, B. Koo, and C. J. Kim, "Evaluation of anodic Ta_2O_5 as the dielectric layer for EWOD devices", *IEEE MEMS*, pp. 428–431, 2012.
17. K. Bohringer, "Modeling and controlling parallel tasks in droplet-based microfluidic systems", *IEEE Transactions on Computer-Aided Design of Integrated Circuits & Systems*, vol. 2, pp. 329–339, 2006.
18. P.-H. Yuh, C.-L. Yang, and Y.-W. Chang, "BioRoute: a network-flow based routing algorithm for digital microfluidic biochips", *Proc. IEEE/ACM International Conference on Computer-Aided Design*, pp. 752–757, 2007.
19. M. Cho and D. Pan, "A high-performance droplet-routing algorithm for digital microfluidic biochips", *IEEE Transactions on Computer-Aided Design of Integrated Circuits & Systems*, vol. 27, pp. 1714–1724, 2008.
20. T.-W. Huang and T.-Y. Ho, "A fast routability- and performance-driven droplet-routing algorithm for digital microfluidic biochips", *Proc. International Conference on Computer Design*, pp. 445–450, 2009.
21. Y. Luo, K. Chakrabarty, and T.-Y. Ho, "Dictionary-based error recovery in cyberphysical digital-microfluidic biochips", *Proc. IEEE/ACM International Conference on Computer-Aided Design*, pp. 369–376, 2012.
22. Y. Luo, K. Chakrabarty, and T.-Y. Ho, "A cyberphysical synthesis approach for error recovery in digital microfluidic biochips", *IEEE/ACM Design, Automation and Test in Europe Conference*, pp. 1239–1244, 2012.
23. V. Kumar and N. Sharma, "SU-8 as hydrophobic and dielectric thin film in electrowetting-on-dielectric based microfluidics device", *Journal of Nanotechnology*, pp. 1–6, 2012.
24. F. Su and K. Chakrabarty, "Unified high-level synthesis and module-placement for defect-tolerant microfluidic biochips", *Proc. IEEE/ACM Design Automation Conference*, pp. 825–830, 2005.

Chapter 5
Optimization of On-Chip Polymerase Chain Reaction

5.1 Introduction

The amount of Deoxyribonucleic acid (DNA) strands available in the biological sample is a major limitation for many genomic bioanalyses [1–4]. For example, in the study of gene dosage in tumor DNA by comparative genomic hybridization, the analysis procedure requires several hundred nanograms of DNA strands for fluorescent labeling [2]. It is difficult to obtain such a large amount of DNA strands directly from biological samples. To overcome this problem, the polymerase chain reaction (PCR) technique is used as the first step for these bioanalyses to amplify (replicate) the initial DNA strands [2, 3, 5]. Based on the categories of operations involved, the procedure of genomic analysis can be divided into three separate stages [6]. The first stage is sample preparation for PCR; the second stage is amplification of the DNA strands; after DNA amplification, the third stage includes the subsequent operations, such as mixing the droplet that contains DNA strands with other reagent droplets, detecting the concentration of intermediate product droplets, and the hybridization of the amplified DNA sequences [4].

Digital microfluidic biochips (DMFBs) are used extensively for the quantitative analysis of biomolecular interactions and they offer a viable platform for performing the three stages of the PCR procedure on the same chip layout [3]. On a DMFB, nanoliter or picoliter-scale droplets can be manipulated by a two-dimensional electrode array using the electrowetting-on-dielectric effect [3, 7, 8]. With sensors (such as photodetectors and a fluorescent microscope) integrated in the microfluidic system, precise control for droplet volume as well as the reaction time for each step in the experiment can be achieved [3, 9]. Compared to conventional instruments and analyzers, the DMFB platform offers several advantages for implementing PCR. It can implement the complete PCR procedure seamlessly, and achieve low reagent consumption, short time-to-results, rapid heating/cooling rates, and integration of multiple processing modules [5, 10]. The size and power consumption of the entire PCR system can also be reduced.

© Springer International Publishing Switzerland 2015

Y. Luo et al., *Hardware/Software Co-Design and Optimization for Cyberphysical Integration in Digital Microfluidic Biochips*, DOI 10.1007/978-3-319-09006-1_5

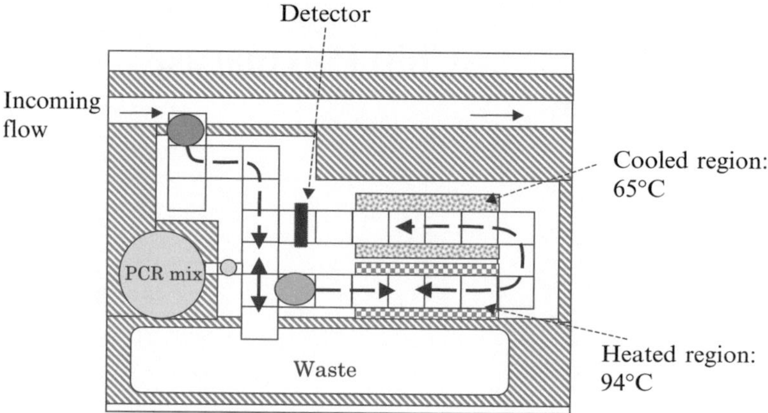

Fig. 5.1 Schematic of a DMFB that can perform DNA amplification [3, 7]

An example of a DMFB performing DNA amplification (i.e., the first stage of the PCR procedure) is shown in Fig. 5.1 [3, 7]. A "heated" region and a "cooled" region are created by a heater. Typically, the temperatures of the heated region and cooled region are set as 94° and 65°, respectively.

First, "PCR mix", which contains Taq DNA polymerase, dNTPs, $MgCl_2$, and reaction buffers in the solution, is prepared on the biochip and stored in a reservoir. Next, the "PCR mix" droplets are blended with the biological sample that contains the target DNA strands. Then, the well-mixed droplet is allowed to pass alternately through several heating and cooling steps [3, 7]. Each thermal cycle contains two steps, i.e., (1) heating the droplet, which splits the double-stranded DNA inside the droplet into two strands, called "DNA melting"; and (2) cooling the droplet, with each strand of the DNA working as a template for synthesizing new DNA strands called "primer annealing". In this way, the initial DNA strands can be amplified into millions of copies [3]. As the amplified DNA strands in the droplet are labeled with fluorophores, the generation rate of the target DNA strands can be measured at each PCR cycle by measuring their intensity of the fluorescence with the sensors integrated in the microfluidic system [4–6].

In order to enable the on-chip hybridization of amplified DNA, some biochips have detection electrodes (DEs) integrated into them [11–14]. Gold detection spots with different immobilized DNA probes are fabricated on the surface of these DEs, as shown in Fig. 5.2 [11, 13]. After the PCR procedure, the droplets that contain the amplified DNA strands are left stationary on the DEs. DNA hybridization occurs if the amplified DNA strands and the probes immobilized on the detection spot are complementary strands. In this method, the DNA sequences in the amplified DNA strands can be analyzed.

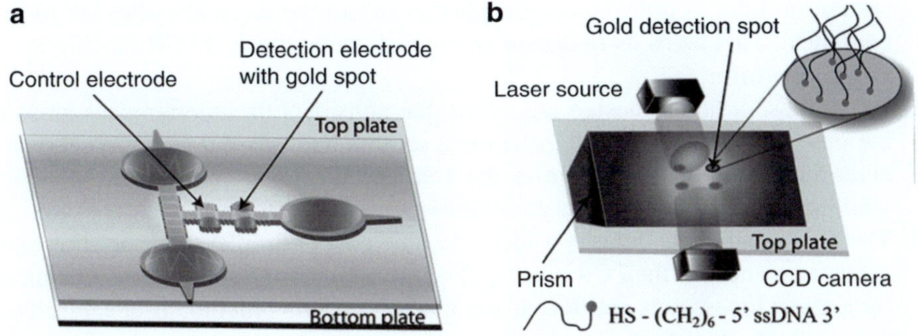

Fig. 5.2 (**a**) Three-dimensional view with the bottom and top plates aligned for a biochip with integrated detection electrodes; (**b**) view of the gold detection spot with DNA attached [11, 13]

Despite offering numerous benefits, prior work on PCR implementations on a DMFB suffers from the following four shortcomings:

1. In all prior work, the functional modules used in the three stages of the PCR procedure are designed separately [3, 7], and they cannot be efficiently integrated on the same layout. This mismatch of the three steps often restricts the effectiveness and feasibility of execution of the complete bioassay.
2. The inherent randomness and complexity of bioanalyses are not considered. When a droplet is dispensed into the biochip, there is a probability that the droplet may not contain sufficient amount of DNA for the PCR (referred to as an "empty droplet") [15]. The biochips designed in prior work are unable to monitor the presence of empty droplets; hence they cannot terminate the wasteful execution of the PCR.
3. The interferences (electrical, thermal, optical, fluidic) among the devices on the biochip, which arise due to proximity effects, are not considered in the previous work. For example, high temperature around the heater may lead to degeneration of a biological sample in a reservoir [16]; therefore, the heater and the reservoir cannot be placed too close to each other.
4. The previously designed PCR biochips are oblivious to the schedule of mixing operations [6]. This may lead to excessive transportation of droplets during mixing/dilution/detection operations in the PCR procedure, which may adversely affect the performance and speed of the biochip.

To overcome the above shortcomings, we propose the first design method that can optimize the PCR procedure on cyberphysical digital microfluidic biochips. The key contributions of this paper are as follows:

1. To overcome the empty-droplet recognition problem, a statistical model for DNA amplification is used to predict whether the droplet has sufficient amount of DNA strands to carry out the PCR. The intensity of fluorescence detected by the photodetectors (PDs) [17–19] or fluorescent microscope are fed back to the

system on-line, in order to decide whether or not the thermal cycles are to be continued. This helps us to design an efficient cyberphysical PCR biochip on a DMFB platform.

2. We propose a layout design algorithm that considers the interferences among the on-chip elements include reservoirs, sensors and thermal units. A heuristic algorithm is applied to minimize the area and electrode count of the biochip while satisfying the proximity constraints.

3. The placement of modules considers the time cost of droplet transportation and defect tolerance for the PCR biochip. The overall mixing/dilution/detection time for a given bioassay is optimized, and a droplet-routing method is developed to bypass electrode defects.

4. The visibility of droplets in the sensing system for cyberphysical biochips is considered. The mixing operations are scheduled to ensure that the sensing system can perform real-time observations of the droplets from both top and side views.

The remainder of this chapter is organized as follows. Section 5.2 describes the statistical model for the amplification of DNA strands, and the on-line decision making method during an actual experiment. Section 5.3 introduces the device placement algorithm with the consideration of device interferences. Section 5.4 describes an application-specific reservoir allocation method. The consideration of droplet visibility during operation scheduling is discussed in Sect. 5.5. Simulation results for three widely used bioassays are presented in Sect. 5.6. Section 5.7 concludes the chapter.

5.2   Cyberphysical Biochip with On-line Decision Making

Figure 2.5 presents the laboratory setup for cyberphysical PCR biochip [9]. The sensing system on biochip monitors the intensity of fluorescence in the droplet, and provides input to the control software. In this way, the software is able to control the execution of the PCR procedure based on the sensor feedback.

When a DNA-binding dye binds to a double-stranded DNA in PCR, it causes fluorescence of the dye. An increase in the DNA product during PCR therefore leads to an increase in fluorescence intensity [4–6]. The sensing system monitors the intensity of fluorescence in the droplet, and provides input to the control software. In this way, the software can control the execution of the PCR procedure based on sensor feedback.

5.2.1   *Statistical Model for the Number of DNA Strands in a Droplet*

When droplets are dispensed from the reservoir into the DMFB, the number of DNA strands contained in a droplet can be considered to be a random variable [15, 20]. Based on the statistical data derived from several biochemistry experiments, we note that the dispensing process follows Poisson statistics when the DNA strands in the biological sample has low density. Therefore, the number of DNA strands in a droplet is derived as [15, 20]:

$$P(X_c = k) \approx e^{-\lambda} \frac{\lambda^k}{k!} \tag{5.1}$$

where k is the number of DNA strands in the droplet and λ is the average number of DNA strands per droplet [15, 20].

When the number of DNA strands in a droplet reaches a low threshold, the PCR procedure cannot be carried out successfully [21]. The droplets that contain low amounts of DNA strands are referred to as "empty droplets". If empty droplets are detected at the end of the PCR, the time spent on running the thermal cycles is wasted. In order to investigate this problem, the feedback signal from the sensors, which are incorporated on the biochip, can be used for making an online probabilistic decision whether a droplet is empty. As noted in [6, 7], during the execution of the PCR, the intensity of fluorescence in the droplets can be monitored by on-chip detectors. The droplet will be discarded if the probability that "the droplet is empty" is high. A new droplet will then be dispensed into the biochip for PCR processing.

In the next part of this section, we introduce statistical models of the PCR bioassay based on which an on-line decision making system will be designed.

5.2.2   *A Simplified Statistical Model for Amplification of DNA*

During the calibration of the dispensing operation, the probability of generating a "good droplet" can be derived [15]. A good droplet is one that contains a sufficient number of DNA strands for running the PCR. Let G denote the event "a good droplet is dispensed into the biochip". The complement of G is the event G^c, i.e., "an empty droplet is dispensed into the biochip".

We assume that when running the PCR on a good droplet, the probability of detecting a fluorescence signal (which indicates that the DNA has been amplified) at the ith thermal cycle is p_i. Let A_k denote the event "no signal is obtained from the droplet at the kth thermal cycle". Therefore, the joint probability of G and A_k can be written as:

$$P(G \cap A_k) = \prod_{i=1}^{k}(1 - p_i) \cdot P(G) \tag{5.2}$$

If no signal is detected at the kth thermal cycle, the conditional probability that "this is a good droplet" (written as $P(G \mid A_k)$) is defined as the quotient of the joint probability of G and A_k, and the probability of A_k:

$$\begin{aligned} P(G \mid A_k) &= \frac{P(G \cap A_k)}{P(A_k)} \\[2mm] &= \frac{P(G \cap A_k)}{P(G \cap A_k) + P(G^c \cap A_k)} \end{aligned} \tag{5.3}$$

where $P(G \cap A_k) = \prod_{i=1}^{k}(1 - p_i) \cdot P(G)$, and $P(G^c \cap A_k) = P(G^c) = 1 - P(G)$. Therefore, we have:

$$P(G \mid A_k) = \frac{\prod_{i=1}^{k}(1 - p_i) \cdot P(G)}{\prod_{i=1}^{k}(1 - p_i) \cdot P(G) + 1 - P(G)}$$

$$P(G^c \mid A_k) = \frac{1 - P(G)}{\prod_{i=1}^{k}(1 - p_i) \cdot P(G) + 1 - P(G)}.$$

5.2.3   An Improved Statistical Model for Amplification of DNA

During the calibration of DNA amplification, one may categorize the droplets based on the number of DNA strands therein. Equation (5.3) remains valid in this scenario.

Let event G_i denote "a droplet with i DNA strands is dispensed into the biochip". We assume that $M_{\min}$ is the minimum number of DNA strands in a droplet needed to run DNA amplification successfully. Then the event G in (5.3) can be expressed as:

$$G = G_{M_{\min}} \cup G_{M_{\min}+1} \cup G_{M_{\min}+2} \cup \cdots \cup G_{\infty}, \tag{5.4}$$

where the events $G_{M_{\min}}, G_{M_{\min}+1}, \cdots, G_{\infty}$ are mutually exclusive.

The event A_k in (5.3) can now be written as:

$$A_k = A_k^0 \cup A_k^1 \cup A_k^2 \cup \cdots A_k^j \cup \cdots \cup A_k^{\infty}, \tag{5.5}$$

where A_k^j is the event that "no signal has been observed from a good droplet with j DNA strands ($j \geq M_{\min}$) after k thermal cycles". It is easy to note that the events $A_k^0, A_k^1, \cdots, A_k^\infty$ are mutually exclusive.

From the definitions of G_i and A_k^j, it follows that $P(G_i \cap A_k^j) = 0$, if $i \neq j$ or $i = j < M_{\min}$. When $i = j \geq M_{\min}$, the value of $P(G_j \cap A_k^j)$ can be empirically derived from calibrating the amplification of DNA. The probabilities $P(G \cap A_k)$ and $P(G^c \cap A_k)$ in (5.3) can be calculated as follows:

$$P(G \cap A_k) = P(G \cap (A_k^0 \cup A_k^1 \cup A_k^2 \cup \cdots A_k^j \cdots \cup A_k^M))$$

$$= P(G_{M_{\min}} \cap A_k^{M_{\min}}) + P(G_{M_{\min}+1} \cap A_k^{M_{\min}+1})$$

$$+ P(G_{M_{\min}+2} \cap A_k^{M_{\min}+2}) + \cdots + P(G_\infty \cap A_k^\infty)$$

$$P(G^c \cap A_k) = P(G_0) + \cdots + P(G_{M_{\min}-1}).$$

We assume that the probability of detecting the fluorescence signal (i.e., the DNA has been amplified) is p_i^m at the ith thermal cycle when we run DNA amplification on a good droplet that contains m ($m \geq M_{\min}$) DNA strands. Then the value of $P(G \cap A_k)$ and $P(G^c \cap A_k)$ can be derived in terms of p_i^m. Therefore, if the fluorescence signal is not detected at the kth thermal cycle, the conditional probability that "the droplet is an empty droplet" as well as "the droplet is a good droplet" can be analyzed in a similar way as the method presented in Sect. 5.2.2.

During DNA amplification on a cyberphysical PCR biochip, the control software calculates the values of $P(G \cap A_k)$ and $P(G^c \cap A_k)$ according to the feedback from the sensor at each thermal cycle. When the value of $P(G^c \cap A_k)$ reaches a threshold (e.g., 95 %), the control software will conclude that the droplet is empty. Then it will initiate a new set of electrode-actuation sequences to discard the empty droplet and dispense a new droplet into the biochip. The PCR procedure will therefore be continued in an adaptive manner on the biochip.

5.3   Optimized Resource Placement Under Proximity Constraints

In this section, we present our algorithm for optimizing the layout under design constraints. The device-proximity constraints for the placement of the devices on the PCR biochip are introduced in Sect. 5.3.1, and the objective function for this placement is presented in Sect. 5.3.2. The geometry-based device placement algorithm is discussed in Sect. 5.3.3. Finally, Sect. 5.3.4 introduces two approaches that can be used to select the optimal result of device-placement.

5.3.1 Device-Proximity Constraints on a PCR Biochip

On the PCR biochip, there are three categories of devices: reservoirs, detection electrodes (DEs), and one thermal unit (e.g., a heater). In order to avoid interferences among the on-chip devices, we assume that the following constraints must be satisfied: which enforce minimum (threshold) intra-pair and inter-pair separation distances between the respective device pairs, must be satisfied. These constants are provided as inputs to biochip designers.

5.3.1.1   Separation Constraints

1. Reservoir to reservoir separation: In order to avoid fluidic leakage between two reservoirs, the distance between them should be no less than a threshold L_{RR} [22]. A typical value of L_{RR} is four times of the length of electrodes on the DMFB [23].
2. DE to DE separation: The stationary droplets on the DEs may have undesirable cross-contamination if they are too close together [11]. Therefore, the distance between two DEs should be no less than a minimum value of L_{DD}. A typical value of L_{DD} is four times the length of electrodes on the DMFB [11].
3. Reservoir to DE separation: Fluids loaded in the reservoirs may contain fluorescent labels and thus they may influence the accuracy of the microscope when measuring the fluorescent intensity of the DE [11]. Furthermore, when a DE and a reservoir are placed close to each other, droplets dispensed from the reservoir may lead to contamination on the surface of the DE. Therefore, the spacing between them should be no less than a threshold L_{RD}. A typical value of L_{RD} is five times the length of electrodes on the DMFB [18].
4. Reservoir to heater separation: Since the biological samples and reagents loaded in the reservoirs may undergo degeneration under heating [16], the distance between a reservoir and a heater should be no less than a threshold L_{RH}. A typical value of L_{RH} equals five times the length of electrodes on the DMFB [3].
5. DE to heater separation: Due to the limitation of fabrication process, the heater on the biochip cannot overlap with any DE on the biochip [24]. Hence the distance between a DE and a heater should be no less than the length of electrodes on the DMFB. The spacing condition between a DE and a heater is written as $L_{DH} \geq 1$.

For simplicity, we define an operator $\mathscr{C}$, which is a mapping from a pair of arbitrarily chosen devices (written as d_a and d_b) to the required minimum distance between them:

$$\mathscr{C} : (d_a, d_b) \rightarrow \text{Minimum distance between } d_a \text{ and } d_b.$$

For example, if DE is represented by d_a, and a heater is represented by d_b, then $\mathscr{C}(d_a, d_b) = L_{DH}$.

In order to design the layout for a PCR biochip with devices $d_{1\sim n}$ on it, we must determine the coordinates for all the devices (written as (x_1, y_1), (x_2, y_2), ..., (x_n, y_n)), such that:

$$\sqrt{(x_i - x_j)^2 + (y_i - y_j)^2} \geq \mathscr{C}(d_i, d_j), \quad \forall i, j.$$

5.3.2 Objective Function for Device Placement

The objective function assesses the quality of the results of device placement. We consider the following two metrics in the objective function:

- The area of the PCR biochips. The fabrication cost of the biochip increases with area. For a given device placement G, assume that the coordinates of all of the devices can be written as (x_1, y_1), (x_2, y_2), ..., (x_n, y_n). The area of the biochip can be estimated as follows:

$$Area = [max(x_1, x_2, \ldots, x_n) - min(x_1, x_2, \ldots, x_n)] \\ \times [min(y_1, y_2, \ldots, y_n) - min(y_1, y_2, \ldots, y_n)]. \tag{5.6}$$

- The average of the distances from the heater to the reservoirs on the biochip. As discussed in Sect. 5.3.1, relatively high temperature around the heater may lead to degradation of the samples in the reservoirs. If the constraint concerning the minimum distance between the reservoirs and the heater is not violated, the overheating of the liquid in the reservoirs can be avoided. However, in order to further guarantee the quality of the biochemical samples in the reservoirs, it is preferable to place the reservoirs as far away from the heater as possible. For a given device placement result G, assume that all the N_r reservoirs can be labeled as (x_{r_1}, y_{r_1}), (x_{r_2}, y_{r_2}), ..., $(x_{r_{N_r}}, y_{r_{N_r}})$ in coordinates, and the coordinates of the heater are (x_h, y_h). Then, the average distance between the heater and the reservoirs can be estimated as follows:

$$D_{avg} = \frac{\sum_{i=1}^{N_r} [(x_{r_i} - x_h)^2 + (y_{r_i} - y_h)^2]^{\frac{1}{2}}}{N_r}. \tag{5.7}$$

For a given device placement result G, the objective function for the optimization of biochip design ($F(G)$) can be computed as follows:

$$F(G) = \alpha Area + \frac{\beta}{D_{avg}(G)} \tag{5.8}$$

where α and β are user-specified weight parameters that can be adjusted to count the relative importance of each metric. For two device placement results G_1 and G_2, if $F(G_1) < F(G_2)$, then G_1 will be considered as the better device placement result.

5.3.3   Device Placement on a PCR Biochip

The device placement problem in a biochip has the following two characteristics:

- The layout of a digital microfluidic biochip usually consists of a discrete 2D array of identical electrodes; hence, we can assume that the devices and electrodes are constrained to stay on integer grid points.
- $\mathscr{C}(d_a, d_b)$, the minimum distance between two devices d_a and d_b, depends on the device pair.

Since each pair of devices may require a distinct constraint on separation spacing, the overall search space may become very large and we conjecture that the problem of finding a minimum-area layout is an NP-hard problem [25]. Hence to develop our design tool, a greedy algorithm based on an iterative method is proposed in this section.

We assume that the PCR layout will be built on a "digital plane", which is the set of all points possessing integer coordinates [26]. Mathematically, the digital plane can be represented as $\mathbb{Z}^2$, where $\mathbb{Z}$ is the set of integers. The center of every physical device/electrode is assumed to be placed on a "digital point", i.e., a point in $\mathbb{Z}^2$ [26].

If d_a and d_b stand for two devices to be placed on the chip, we draw a circle with radius $\mathscr{C}(d_a, d_b)$, centered at d_a. The region inside the circle is defined as the "forbidden region" of d_b with respect to d_a. The set of all electrodes that completely or partially overlap with the forbidden region of d_b is written as $O_{(d_a, d_b)}$. Note that $O_{(d_a, d_b)}$ can be considered as a "digital object" in $\mathbb{Z}^2$ as it is a set of digital points [26]. Figure 5.3a shows an example of the digital object corresponding to the heater.

If d_b is placed on an electrode that is not an element of $O_{(d_a, d_b)}$, then the distance between their centers will be no less than $\mathscr{C}(d_a, d_b)$. In other words, the corresponding separation constraint will be satisfied. As shown in Fig. 5.3a, if the reservoir is placed at the boundary or outside of the heater's digital object, the separation constraint for the reservoir and the heater will be satisfied.

In order to minimize the area of the chip, all the devices need to be placed as closely as possible. An example of valid and compact placement of a reservoir R_2 in the presence of a heater H, a DE D_1, and a reservoir R_1 is shown in Fig. 5.3b. The isothetic convex envelope [26] that tightly encloses the union of three forbidden regions (corresponding to H, D_1, and R_1, each with respect to R_2) is displayed here as a digital object. Note that for a valid placement, the reservoir R_2 should not be placed inside this object; also for compactness, it should be placed as closely as possible so that the total area of the chip including all the resources is minimized.

In the initial stage of device placement, all the devices of the PCR biochip are listed in a priority queue, Q_{dev}. The ordering of devices in Q_{dev} is randomly

Fig. 5.3 (**a**) An example of the digital object corresponding to the heater; (**b**) the isothetic convex envelope (*ICE*) that tightly encloses the forbidden region of R_2; optimal placement of a reservoir R_2 in the presence of a reservoir R_1, a DE D_1, and a heater H

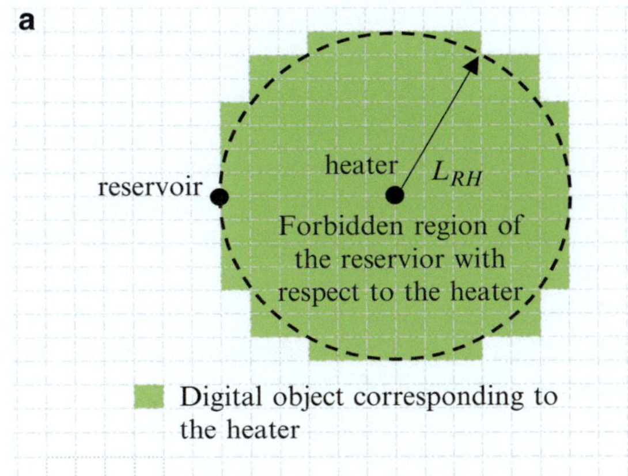

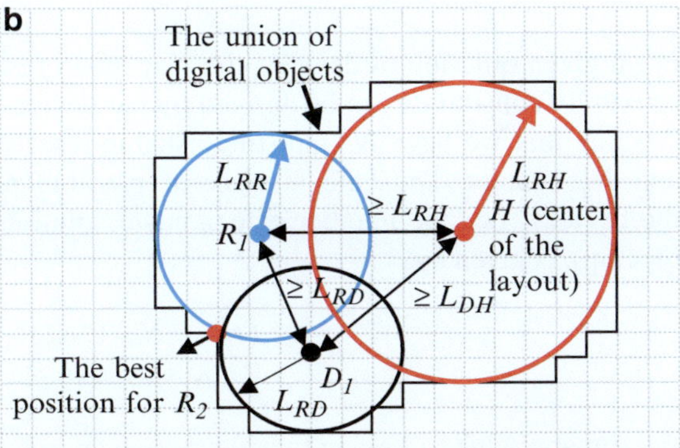

generated. In our iterative method, first we choose the first device in Q_{dev} (which is written as d_{p_1}) and place it at the origin $(0,0)$ of the layout grid. Then, the second device in Q_{dev} (which is written as d_{p_2}) is chosen for placement.

We draw a circle of radius $\mathscr{C}(d_{p_1}, d_{p_2})$ with its center at d_{p_1}. While placing d_{p_2}, we flag all the grid points that lie inside the circle as being in the "forbidden" region for d_{p_2}. The digital object for d_{p_1}, $O_{(d_{p_1}, d_{p_2})}$, can also be derived.

Thus d_{p_2} may be placed on any electrode that is not an element in $O_{(d_{p_1}, d_{p_2})}$ to satisfy the separation constraint. At any instant of time, let us assume that the devices $d_{p_1}, d_{p_2}, \ldots, d_{p_k}$ have already been placed. When considering the placement of the device $d_{p_{k+1}}$, we proceed as follows: for each device d_{p_i} already placed, we draw a forbidden circle centering at d_{p_i} with radius $\mathscr{C}(d_{p_i}, d_{p_{k+1}})$, and derive the corresponding digital object $O_{(d_{p_i}, d_{p_{k+1}})}$. The union of all these digital objects, which is written as $O_{(d_{p_{1\ldots k}}, d_{p_{k+1}})}$, is defined as follows:

$$O_{(d_{p_{1\ldots k}}, d_{p_{k+1}})} = O_{(d_{p_1}, d_{p_{k+1}})} \cup O_{(d_{p_2}, d_{p_{k+1}})} \ldots \cup O_{(d_{p_k}, d_{p_{k+1}})}.$$

The device $d_{p_{k+1}}$ may be placed anywhere outside $O_{(d_{p_{1...k}}, d_{p_{k+1}})}$. However, we place it on a grid point such that the value of $Area(d_{p_{1...k+1}})$ is minimized, where $Area(d_{p_{1...k+1}})$ is the area occupied and contoured by devices $d_{p_1}, d_{p_2}, \ldots, d_{p_k}, d_{p_{k+1}}$. The value of $Area(d_{p_{1...k+1}})$ is calculated according to (5.6).

This procedure of device placement is illustrated in Fig. 5.3b, where in the presence of a heater H, a DE D_1, and a reservoir R_1, we show how a new reservoir R_2 can be optimally placed. We also assume that the heater is placed at the origin of the layout as the first step of our algorithm.

In each subsequent iteration, we select the device from the top of the priority queue Q_{dev}. The device is optimally placed with respect to the resources so far placed in each iteration.

5.3.4 Optimization of Device Placement Results

For a given ordering of devices, the placement results that satisfies all the device-proximity constraints can be derived by the device placer introduced in Sect. 5.3.3. Assume that we have N_r indistinguishable reservoirs, N_d indistinguishable DEs, and N_h heaters. Thus the number of possible priority queues that contain all the devices (i.e., the number of candidate solution for device placement on PCR biochip), is given by:

$$\frac{(N_r + N_d + N_h)!}{N_r! N_d! N_h!}.$$

We can apply the following two alternative approaches in order to select the "best" candidate solution based on the cost function introduced in Sect. 5.3.2.

5.3.4.1 Approach 1: Enumeration of All the Possible Ordering of Devices

In practice, only a few EDs are integrated on a biochip due to their high fabrication costs, and a single heating/cooling module suffices. The number of reagents required in a typical mixing procedure after DNA amplification is around 8. It is customary to place the output ports of reservoirs around the boundary of the chip and the heater in the interior. Therefore, the number of possible orderings significantly reduces in a practical scenario. Hence we can execute the above-mentioned heuristic algorithm exhaustively on all orderings. The value of the cost function introduced in Sect. 5.3.2 can be calculated for the placement of each device. The placement that corresponds to the minimum value of the cost function is selected as an "optimal solution".

5.3.4.2 Approach 2: Simulated Annealing-Based Device Placement

The computational complexity of Approach 1 may become prohibitively high as the number of devices on a PCR biochip increases. In this case, we present a genetic algorithm-based (GA-based) approach to search for an optimal placement of the devices.

In the GA-based approach, each priority queue that contains all the devices on the PCR biochip is encoded as a "chromosome". A chromosome is a vector of random keys (or "genes") defined as below:

$$Chromosome = (gene(1),\ gene(2),\ gene(3),\ \dots, gene(n)),$$

where n is the number of devices that must be placed, and $gene(i)$ ($i = 1$ to n) is a number sampled from $[0, 1]$. The priority value of device i is set as: $P_{dev}(i) = gene(i)$.

By sorting the priority values, a priority queue listing all the devices can be derived. Then using the device placer introduced in Sect. 5.3.3, a feasible placement that satisfies the proximity constraints can be constructed by using any chromosome.

At the beginning of the GA-based approach, $4n$ chromosomes are randomly generated. The following evolutionary operators are considered in the heuristic approach:

- *Reproduction.* The chromosomes that correspond to the minimum value of the cost function are copied to the next generation.
- *Crossover.* In the crossover procedure, two parent chromosomes are randomly chosen from the old population. Their offspring chromosome is generated in the following way: $gene(i)$ of the offspring chromosome can either be copied from $gene(i)$ of the parent chromosome. The probability that $gene(i)$ is inherited from the father and mother chromosomes are set as P and $1 - P$, respectively. The offspring chromosome is added into the new population of the next generation.
- *Mutation.* The new chromosomes of the population are randomly generated to guarantee the diversity of the population.

In the GA-based heuristic, we keep the number of chromosomes in each generation to be $4n$. During the evolution, the n best chromosomes (i.e., the chromosomes that correspond to the n minimum values of the cost function) are reproduced into the next generation. A total of $2n$ chromosomes in the new population are the offspring generated from the previous population, and n new chromosomes are randomly generated. This proportion can be fine-tuned further through experiments. After several generations of evolution, the chromosome with the minimum value of cost function is selected from the final population. The solution for the placement of the devices is constructed by using the chosen chromosome. In Sect. 5.4, we will discuss how to optimize droplet transportation and mixing time in the context of a specific PCR protocol.

5.4   Bioassay-Specific Reservoir Allocation for PCR Biochip

In a typical mixing procedure after DNA amplification, several reagents/samples are required to be mixed in a certain ratio. The number of mix-split operations depends on the mixing algorithm and the target ratio. The mixing process is generally represented by a mixing (or a sequencing) graph G. Figures 3.10 and 5.4 shows two examples of the sequencing graphs for solution preparation procedures [27, 28]. The sequencing graph of a bioassay is a directed graph. Each node of zero in-degree in the graph represents a dispensing operation, and each internal node represents a fluidic operation (e.g., mixing/dilution) [27]. The droplet transportation cost in the PCR should be optimized based on the given sequencing graph of the bioassay, which requires appropriate allocation of reservoirs to the reagents and scheduling of mixing/dilution operations as specified in the sequencing graph.

In this section, we will discuss the layout-design algorithm for PCR biochip. The "electrode ring" structure for low-cost biochip is introduced in Sect. 5.4.1. The droplet routing problem on low-cost PCR biochip is introduced in Sect. 5.4.2. Section 5.4.3 presents the algorithm of bioassay-specific resource allocation for PCR biochip.

5.4.1   Electrode Ring on Low-Cost Biochips

The placement algorithm described in Sect. 5.3.3 assigns a location for each of the devices on a 2D grid. As mentioned in Sect. 5.3.3, a heater is usually placed at the center of the layout; reservoirs are placed on the boundary of the biochips. In order to complete the layout design of the PCR biochip, the positions of electrodes on the biochips have to be determined, and droplet transportation paths among the devices need to be created.

In order to reduce the number of electrodes in a typical low-cost biochip, the output ports of all reservoirs on the layout are connected in a "cycle" consisting

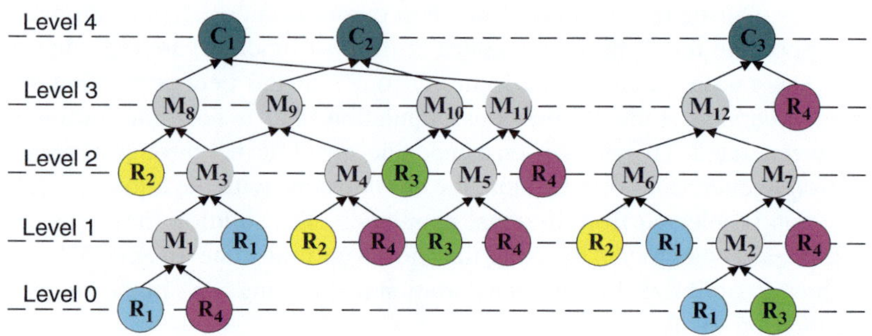

Fig. 5.4   An example of the sequencing graph [28]

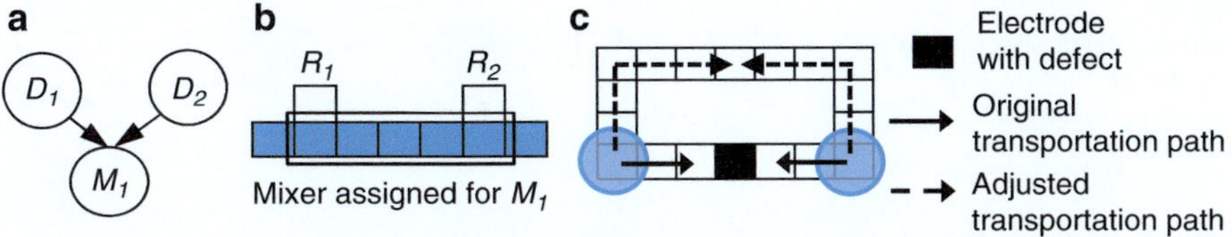

Fig. 5.5 (**a**) Sequencing graph for the dispensing and mixing of two droplets; (**b**) placement of the output port of reservoirs and the mixer; (**c**) an example of defect tolerance by droplet rerouting

of electrodes [7, 8]. There is no pre-assigned specific region for mixer modules; all the mixing/splitting operations are performed on the electrodes lying between the output ports of reservoirs. Let D_1 and D_2 denote two dispensing operations from reservoirs R_1 and R_2 respectively, and let M_1 represent a mixing operation between these two droplets (Fig. 5.5a). Figure 5.5b shows a possible placement of these modules, in which the droplets dispensed from R_1 and R_2 are mixed on the electrode array between these two reservoirs.

By placing the output of reservoirs on the circular electrode path, the defect tolerance of PCR biochips can be enhanced. An example is shown in Fig. 5.5c. Two droplets are scheduled to be mixed together, and their original transportation paths are indicated by the arrows. When a defect is present on the layout, the transportation paths of these two droplets are dynamically adjusted. Therefore, using an alternative droplet transportation path, the PCR biochip can execute bioassays despite the presence of the defect.

5.4.2 Droplet Routing on Low-Cost PCR Biochip

In Sect. 5.3.3, when determining the placement of devices, all the reservoirs are considered as indistinguishable devices. However, in order to execute a given bioassay on the PCR biochip, we have to allocate reservoirs for each sample/reagent used in the bioassay.

When multiple operations need to be concurrently performed on a low-cost biochip with "electrode ring" structure, the routing paths of droplets may have conflicts. An example is shown in Fig. 5.6. Here we assume that reagents x_1 to x_7 shown in Fig. 3.10 are stored in reservoirs R_1 to R_2.

As defined by the sequencing graph (Fig. 3.10), the droplets dispensed from reservoirs R_1 and R_4 need to be mixed together in operation M_5, and the droplets dispensed from R_3 and R_5 need to be mixed together in operation M_3. We assume that the positions of outputs for R_1, R_3, R_4, and R_5 are placed like Fig. 5.6. In order to perform mixing operation M_3 and M_5, the droplets dispensed from R_1, R_3, R_4, and R_5 need to be transported in the directions indicated by the arrows in Fig. 5.6. It is easy to observe that in order to avoid the head-on collisions, these four droplets cannot be transported on the biochip concurrently. Hence the mixing operations M_3 and M_5 cannot be executed in parallel.

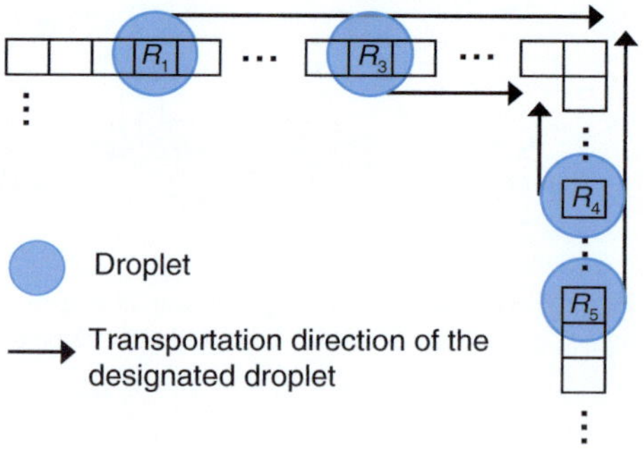

Fig. 5.6 Conflicts of droplet routing. Mixing operations M_3 and M_5 in Fig. 3.10 cannot be executed concurrently

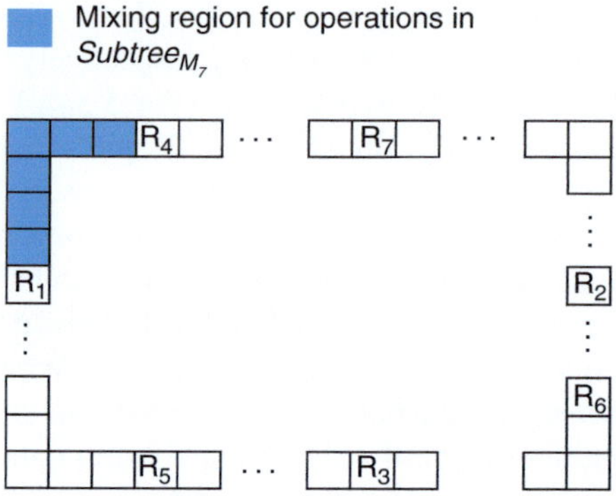

Fig. 5.7 The placement of reservoirs that can reduce conflicts of droplet routing

The conflicts of droplet routing can be reduced by changing the configuration of reservoir allocation. Assume that the outputs of reservoirs $R_1 \sim R_7$ are placed as shown in Fig. 5.7, the degree of parallelism for operation execution will increase. As shown in Fig. 3.10, the operations in the subtree with root node M_9 (which is written as $Subtree_{M_9}$) only involves droplets dispensed from reservoirs R_1 and R_4. Therefore all the operations of $Subtree_{M_9}$ can be performed on the electrodes that are directly connected with the output ports of R_1 and R_4, as shown in Fig. 5.7. As there is no output port of other reservoir placed in this region, droplets dispensing from R_1 and R_4 can be mixed without conflicts of droplet routing.

Similarly, operations in $Subtree_{M_8}$, $Subtree_{M_{13}}$, and $Subtree_{M_{14}}$ can be performed in the regions listed in Table 5.1, where the set of electrodes that directly connects the output ports of reservoirs R_i, R_j, ..., R_k is written as $Region(R_i, R_j, \ldots, R_k)$. Hence all the mixing operations in the mixing tree can be performed "locally", and no conflict of droplet routing occurs.

Table 5.1 Operations and their corresponding regions to be performed

Operations	Corresponding region to be performed
$Subtree_{M_8}$	$Region(R_4, R_7)$
$Subtree_{M_9}$	$Region(R_1, R_4)$
$Subtree_{M_{13}}$	$Region(R_2, R_6)$
$Subtree_{M_{14}}$	$Region(R_3, R_5)$
M_{12}	$Region(R_4, R_7)$
M_{15}	$Region(R_4, R_7)$
M_{16}	$Region(R_3, R_6)$
M_{17}	$Region(R_4, R_7, R_2, R_6)$
M_{18}	$Region(R_2, R_7)$

5.4.3   Bioassay-Specific Reservoir Allocation

Based on above discussion, the results for bioassay-specific reservoir allocation, droplet routing, and the scheduling of fluid-handling operations can be derived as follows.

1. **Clustering of sequencing graph**: As discussed in Sect. 5.4.2, if the operations in a subgraph $SubG$ only involve droplets from two reservoirs, then all the operations in $SubG$ can be performed in the region between these two reservoirs. Therefore, in this stage, we need to find the maximal subgraph in the sequencing graph, such that the operations in the subgraph involve only two reagents/samples RS_i and RS_j. This subgraph is defined as "basic mixing clusters" (written as $cluster(RS_i, RS_j)$) in the sequencing graph.

   Starting from the nodes that represent the final mixing operations of the bioassay (i.e., the nodes of zero out-degrees), we check the number of "components" of its output droplets. For example, the output droplet of operation M_18 is a mixture of four reagents R_1, R_2, R_3, and R_4. Then we need to trace back to the predecessor operations of M_18, and then check the components of their output droplets. This trace-back procedure will repeat until finding the operation N^* that the corresponding output droplets have no more than two components. Then the subgraph consisting of operation N^* and its predecessor operations is a basic mixing cluster in the sequencing graph.

2. **Reservoir allocation:** After finding out all the basic mixing clusters of a sequencing graph, the bioassay-specific reservoir allocation can be determined. For each mixing cluster $cluster(RS_i, RS_j)$, the reservoirs corresponding to RS_i and RS_j will be placed as the "neighbors" on the layout. The concept of neighbors on the electrode ring is defined as below:

Definition 1.   Assume that there is a set of reservoirs $S_R = \{R_{n_1}, R_{n_2}, R_{n_3}, \ldots, R_{n_k}\}$, and all these reservoirs are placed on an electrode ring. A pair of reservoirs R_{n_i} and R_{n_j} ($n_1 \leq n_i < n_j \leq n_k$) are "neighbors" in S_R if there is an electrode path $P_{(R_{n_i}, R_{n_j})}$ goes through the output ports of R_i and R_j, and there is no output port of other reservoirs contained by the set S_R on this electrode path.

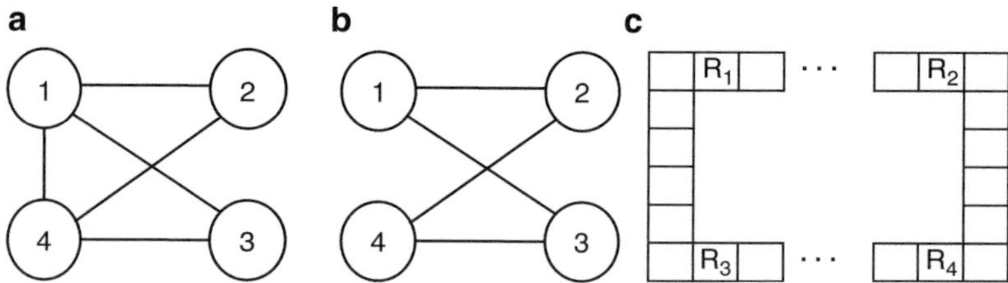

Fig. 5.8 (**a**) The undirected graph that represents the relationship among the basic mixing clusters in Fig. 5.4; (**b**) the adjusted graph after removing the edge between nodes *1* and *4*; (**c**) the ordering of reservoirs on the electrode ring that can avoid conflicts of droplet routing

Definition 2. If going from R_{n_i} to R_{n_j} along the electrode path $P_{(R_{n_i}, R_{n_j})}$ is in clockwise direction, then R_{n_j} is the "clockwise neighbor" of R_{n_i}. Otherwise, R_{n_j} is the "anti-clockwise neighbor" of R_{n_i}.

An example is as follows. The sequencing graph in Fig. 5.4 can be divided into five basic mixing clusters: $cluster(R_1, R_4)$, $cluster(R_2, R_4)$, $cluster(R_3, R_4)$, $cluster(R_1, R_2)$, and $cluster(R_1, R_3)$. The relationship among these mixing clusters can be denoted by an undirected graph. Each node in the graph represents a reservoir; if two reservoirs are in the same cluster, then an edge is added between the pair of nodes that correspond to the reservoirs. The undirected graph corresponding to Fig. 5.4 is shown in Fig. 5.8a. As on the electrode ring of low-cost biochips, each droplet has only two possible routing directions. If a reservoir belongs to more than two basic mixing clusters, droplets dispensed from the reservoir cannot be used in the mixing operations of all these clusters concurrently. Accordingly, in the undirected graph, part of the edges will be removed such that each node has no more than two edges. We have the following conjecture:

Conjecture 5.1. Assume that G_C is a graph that represents the relationship of mixing clusters in a bioassay, and each node in G_C has no more than two edges. Then we can always place the reservoirs on the electrode ring such that every reservoir pair that has an edge between them can be placed as neighbors.

For two reservoirs R_{n_i} and R_{n_j}, if there is an edge between them, then we place R_{n_i} as the neighbor of R_{n_j}. As discussed above, each reservoir on the electrode ring has two neighbors: clockwise neighbor and anti-clockwise neighbor. If one of the neighbor positions (e.g., clockwise neighbor) of R_{n_j} is occupied by another reservoir, then we place R_{n_i} to the unoccupied position. Note that each edge has no more than two edges, hence it will never happen that both of the neighbor positions of R_{n_j} are occupied by other reservoirs.

3. **Scheduling of operations:** Once the reservoir allocation step is completed, the scheduling of operations can be determined based on the sequencing graph. As operations in each basic mixing cluster can be performed locally without long distance transportation, operations in different clusters can be performed

---

1: **Phase1**:
2: **for** each permutation of n devices (written as $\{d_{k1}, d_{k2}, ..., d_{kn}\}$) to be placed on the layout
   do
3:     Place d_{k1} at the origin of the layout (usually, d_{k1} represents the heater));
4:     **for** each $d_{ki} \in \{d_{k2}, d_{k3}, ..., d_{kn}\}$ **do**
5:         Draw the forbidden circular regions constrained by each of the devices $\{d_{k1}, d_{k2},$
           $..., d_{k_{i-1}}\}$ with respect to d_{ki}. Construct the union of digital objects enclosing the
           forbidden regions;
6:         Find the grid point g_i on the boundary of the union of digital objects, such that the
           distance between the origin grid and device d_{ki} is minimum.
7:     **end for**
8:     Output the placement geometry and calculate the cost function;
9: **end for**
10: Choose the device placement that can minimize the value of the cost function; // At this
    step, the locations of all the devices are determined. All the reservoirs are considered iden-
    tical.
11: **Phase 2**:
12: Find out the basic mixing clusters in the sequencing graph;
13: Determine the best reservoir allocation of reagents based on the clustering result of the
    sequencing graph;
14: Establish routing paths to connect all the devices following the sequencing tree;
15: Schedule operations included in basic mixing clusters;
16: Schedule operations not included in basic mixing clusters;

Fig. 5.9 Pseudocode for layout design for PCR biochips

concurrently. The scheduling of operations in the same cluster can be determined based on the input-output interdependency [29].

Operations included in the basic mixing clusters are initially performed when executing the bioassay. Other operations are executed after the operations in mixing clusters are completed. For example, in Fig. 5.4, operations M_1 and M_3 are in the basic mixing cluster $cluster(R_1, R_4)$. After operations in $cluster(R_1, R_4)$ are completed, one of the output droplets of $cluster(R_1, R_4)$ will be further mixed with droplet R_2 in operation M_8; the output droplets of $cluster(R_1, R_4)$ and $cluster(R_2, R_4)$ will be mixed in operation M_9. In this approach, the scheduling of all the fluid-handling operation in the sequencing graph can be derived.

The pseudocode for the entire design cycle is shown in Fig. 5.9 (here the selection of optimal solution is based on Approach 1).

5.5 Visibility-Aware Droplet Detection

In a cyberphysical digital microfluidic system, the monitoring of droplets is vitally important. Researchers often use cameras and fluorescent microscopes, which are fixed on the hardware platform to simultaneously monitor multiple operations on the biochip [30, 31]. From the captured images, the volumes and concentrations of the droplets can be determined at each step of the bioassay, and the time required to complete the dilution/mixing processes on the biochip can be measured precisely [30, 31].

However, misuse of top view image will lead to inaccurate results concerning the completion time of mixing operations. For example, a fluorescein droplet and a non-fluorescein droplet are merged together for 10 s, and fluorescent microscope capture images of the mixed droplet from top and side views, as shown in Fig. 5.10 [31]. From the top view, it is observed that the fluorescent reagent has been homogeneously distributed in the merged droplet, i.e., the mixing procedure may appear to be complete. However, from the side view, the merged droplet has multiple layers which indicates incomplete mixing operation. Hence it is essential to monitor cyberphysical microfluidic biochips with different perspectives. Two cameras can be fixed onto the hardware platform so that images of the droplets can be acquired from two different perspectives.

When multiple droplets are manipulated concurrently on the biochip, the visibility of the droplets for the side-view assessment should be considered. An example is shown in Fig. 5.11a. We assume that droplets 1 $\sim$ 4 are being mixed concurrently in four mixers on the biochip, and their directions of movement are indicated by the arrows. We assume that the mixing operations for these four droplets start at the same time with the droplets moving in the same direction.

As shown in Fig. 5.11a, the initial positions of droplets 1 and 3 are in the same row of the electrode array, and the initial positions of droplets 2 and 4 are in the same row of the electrode array. When the movements of droplets 1 and 3 are projected on the y-axis, they will always overlap with each other during the mixing procedure. Therefore, if the camera is fixed at Position 1 shown in Fig. 5.11a and we acquire side-view images of these four droplets, droplet 1 will always be hidden behind droplet 3, i.e., droplet 1 is "invisible" to the monitoring system. Likewise, droplet 2 will always be hidden behind droplet 4, and it is also invisible to the monitoring system. Similarly, if the camera is fixed at Position 2 shown in Fig. 5.11a and monitors these four droplets from the side, droplets 2 and 4 will always be invisible to the monitoring system.

The situation that "some of the droplets are invisible for the camera or fluorescent microscope" must be avoided in order to monitor all of the mixing procedures on the biochip simultaneously.

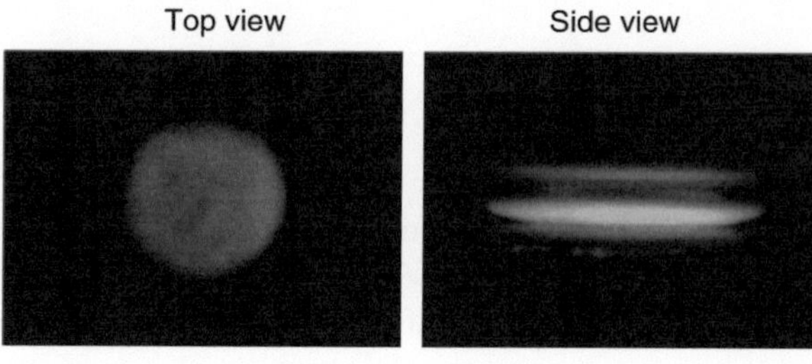

Fig. 5.10 *Top* and *side views* captured after merging a fluorescein droplet with a non-fluorescein droplet for 10 s [31]

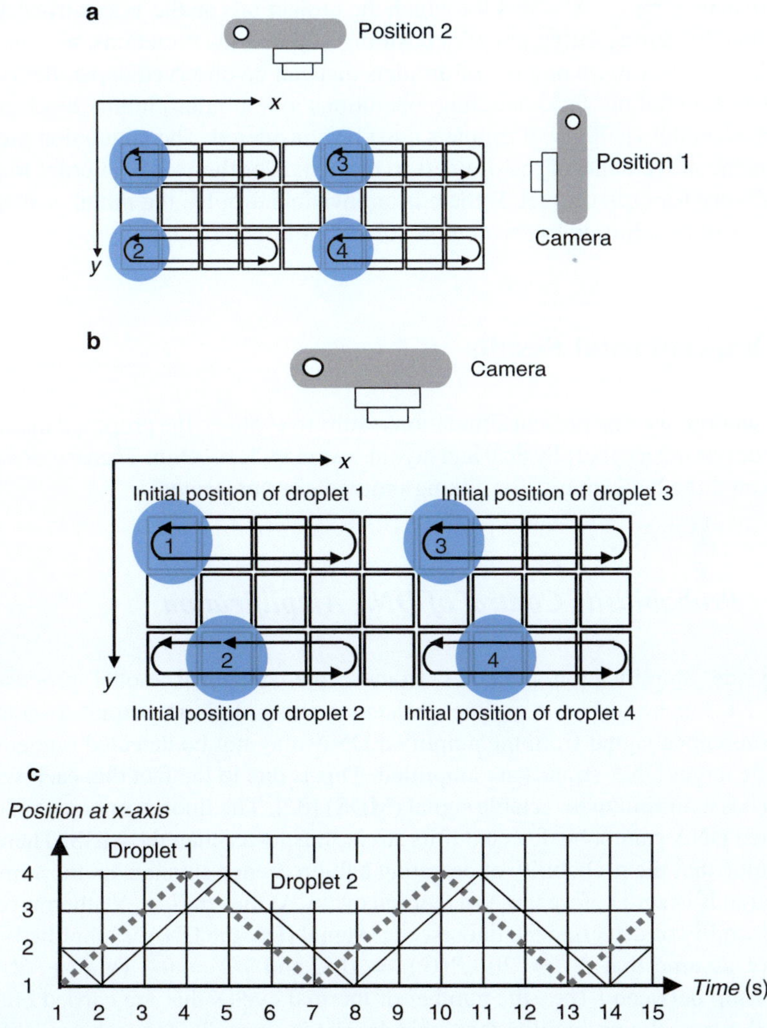

Fig. 5.11 (**a**) Droplets 1 and 2 are invisible from side view if the camera is fixed at Position 1. Droplets 2 and 3 are invisible from side view if the camera is fixed at Position 2; (**b**) set different initial positions for droplets 3 and 4; (**c**) the movements of droplets 1 and 2 in the x direction

To solve this problem, we can set different initial positions for droplets 1 and 2, as shown in Fig. 5.11b. When projecting the movements of droplets 1 and 2 on the x-axis, the relationship between time and the positions of these two droplets is shown in Fig. 5.11c. At each time moment, the positions of droplets 1 and 2 do not overlap each other, so both of these droplets can be photographed by the camera.

By projecting the movements of droplets onto the x or y axis, the maximum number of droplets that can be observed from the side view can be further calculated.

As an instance, for 1×4 mixers for which the projections on the y-axis overlap with each other, by setting different initial positions and moving directions, we can determine that the maximum number of droplets that can be observed in parallel is six.

When scheduling fluid-handling operations on a microfluidic biochip, this constraint on the visibility of droplets must be considered. The simulation program projects the movements of the droplets to the x-axis or the y-axis in order to check the visibility for each droplet. If there is an invisible droplet, the initial positions of droplets will be adjusted by the control software of the biochip.

5.6   Experimental Results

In this section, we first present simulation results to evaluate the proposed methodology of designing a cyberphysical and layout-aware PCR biochip. Then experimental results on three benchmarks for mixing protocols are presented.

5.6.1   Probabilistic Control of DNA Amplification

During the amplification of DNA strands, the statistical model proposed in Sect. 5.2.1 can be used for on-line decision making. It is important to note that the fluorescence signal from the amplified DNA may not be detected immediately when the target DNA strands are amplified. This is due to the fact that each sensing system has a minimum detectable signal (MDS) [32]. The fluorescence signal of the amplified DNA cannot be detected if its strength is lower than the MDS. Therefore, we assume that the probability of detecting a fluorescence signal from the amplified DNA strands is zero before the Nth thermal cycle. At the ith ($i \geq N$) thermal cycle, the probability of detecting the fluorescence signal (i.e., the DNA is amplified) is p_i.

If we assume that $N = 20$, $P(G) = 0.8$, and $p_i = 0.3$ ($\forall\, i \geq 20$), the relationship between i (i.e., the number of thermal cycles that are carried out) and $P(G^c | A_i)$ (i.e., the probability that "this droplet is empty") is shown in Fig. 5.12a. From Fig. 5.12a, we observe that, if there is no sensor signal at the 25th, 27th, or 28th thermal cycles, the value of $P(G^c | A_i)$ is 85%, 90%, or 95%, respectively. The relationship between i and $P(G^c | A_i)$ for different p_i and $P(G)$ are shown in Fig. 5.12a.

When the PCR is performed successively, there is an exponential increase in the number of target DNA strands. We can assume that the probability p_i increases exponentially. For example, we can set $N = 20$, $P(G) = 0.8$, and set p_i as follows:

$$p_i = \begin{cases} 0, & \text{if } i < N \\ p^* \times 2^{i-N}, & \text{if } N \leq i \leq N - \log_2 p^* \\ 1, & \text{if } N - \log_2 p^* < i \end{cases}$$

Fig. 5.12 Relationships between i (i.e., the number of thermal cycles that have been carried out) and $P(G^c|A_i)$ (i.e., the probability that "this droplet is an empty droplet") derived by three statistical models of PCR procedure

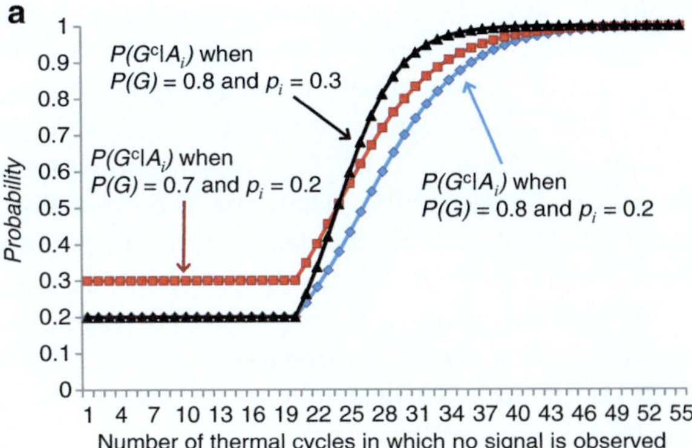

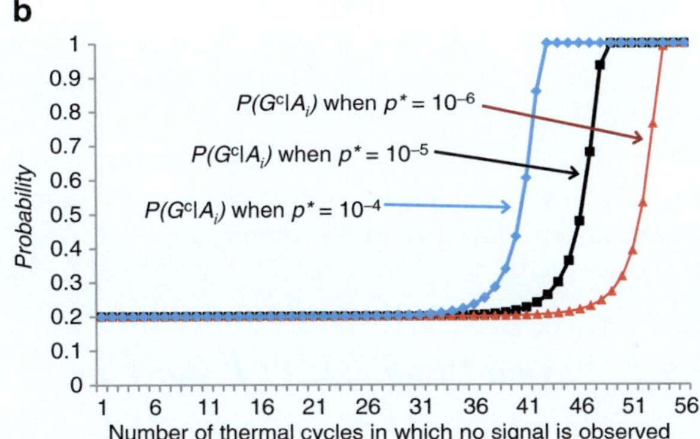

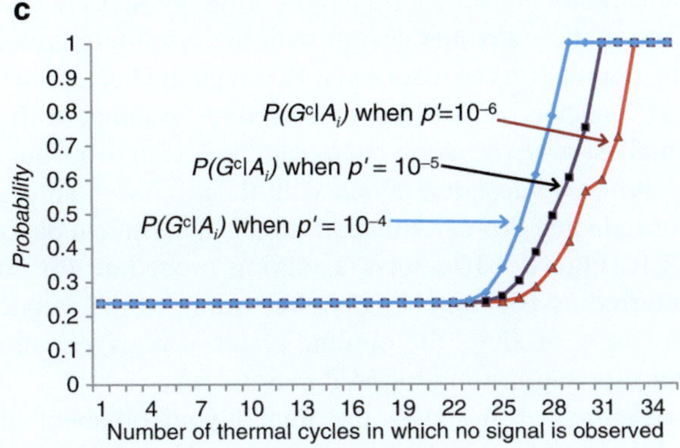

In this scenario, the corresponding relationship between $P(G^c|A_i)$ and i is shown in Fig. 5.12b. In the three curves shown in Fig. 5.12b, the value of p^* is set as 10^{-4}, 10^{-5}, and 10^{-6}, respectively. From the figure we can conclude that when p^* is set as 10^{-5}, if there is no signal at the 36th thermal cycle, the probability that "this droplet is an empty droplet" is 60 %; if there is no signal at the 37th thermal cycle, the probability becomes 99 %. Therefore, if no signal is detected after the 37th thermal cycle, we have a 99 % confidence level that the droplet does not contain enough DNA strands for PCR, and the droplet should be discarded.

We can also analyze the stage of DNA amplification by considering the more detailed statistical model introduced in Sect. 5.2.3. As in the ideal case, the number of DNA strands in a droplet increases exponentially, we assume that the value of p_i^m can be written as follows:

$$p_i^m = \begin{cases} 0, & \text{if } i < N \\ p' \times m^{i-N}, & \text{if } N \leq i \leq N - \log_2 p' \\ 1, & \text{if } N - \log_2 p' < i \end{cases}$$

where $M_{\min} = 3$ and $N = 20$. For the distribution of the number of DNA strands in droplets shown in (5.1), we assume that $\lambda = 4$. Figure 5.12c shows the relationships between i and $P(G \mid A_i)$ for different p^*.

5.6.2 Layout Design for PCR Biochips

Consider a biochip with seven reservoirs, two PDs, and one heater; the separation constraints are set as $L_{RR} = 4$, $L_{DD} = 4$, $L_{RD} = 5$, $L_{RH} = 5$, and $L_{DH} = 1$. One of these layouts design with the minimum area is shown in Fig. 5.13a, where the output ports of reservoirs, PDs, and the heater are represented by "R", "D", and "H", respectively. If each electrode is assumed to have unity area, the size of the smallest-area rectangle enclosing the layout turns out to be 11×11.

After we design a layout with the minimum area, a suitable reservoir allocation for reducing droplet transportation is determined based on the protocol of a specific PCR. Figure 3.10 shows a mixing procedure for solution preparation, which is referred as Bioassay 1 [27]. According to the frequency of reactant usage in the mixing procedure, the optimal result of reservoir allocation can be derived by the heuristic algorithm proposed in Sect. 5.4.

Figure 5.13b shows the final layout obtained after reservoir allocation that minimizes the droplet transportation time. Reagents $x_{1\sim7}$ in Fig. 3.10 are loaded into $R_{1\sim7}$. While executing the bioassay, the electrode path that connects the output ports for a pair of reservoirs can be assigned as mixers. All the mixing operation in the protocol can thus be performed "locally" without long-distance transportation or conflicts in droplet routing. For example, the inputs of operations M_1, M_4, and M_8 are reagents x_4 and x_7, and accordingly, these mixing operations will be performed on the electrodes between R_4 and R_7.

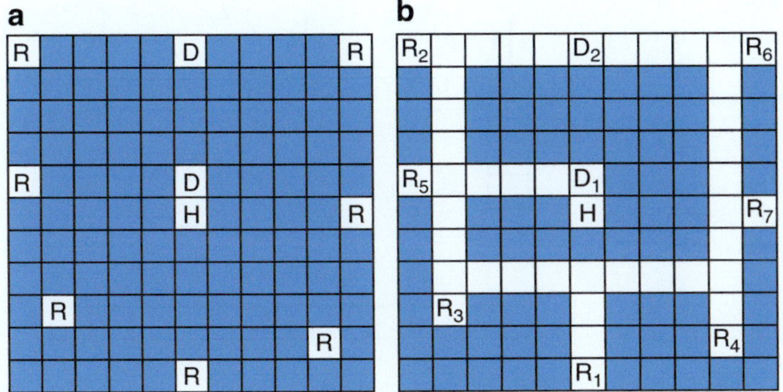

Fig. 5.13 (**a**) Device placement obtained by the proposed algorithm. R represents the output port of the reservoir, D represents the DE, and H represents the heater. All the reservoirs are treated as identical devices and all the PDs are treated as identical devices; (**b**) Reservoir allocation with the minimum droplet routing cost. $R_{1\sim7}$ are reservoirs assigned to inputs $x_{1\sim7}$ in Fig. 3.10, D_1 and D_2 are DEs, and H is the heater

In addition, the degree of parallelism for fluid-handling operation is high since the conflicts of resource-sharing and droplet routing on the biochip can be avoided. Here we assume that the completion time of mixing on a 1×4 mixer is 5 s, and the time of moving the droplet from electrode to another electrode is t_m s (t_m usually varies from 0.01 to 1 [33]). Then the execution time of the whole bioassay on the layout shown in Fig. 5.13b will be calculated as $(35+36t_m)$ seconds.

Here we design another device placement as the baseline algorithm. In the baseline method, all the devices (including the output ports of reservoirs, PDs, and the heater) are placed at the boundary of the layout one by one. The distance between any two devices is set as $L_{max} = \max\{L_{RR}, L_{DD}, L_{RD}, L_{RH}, L_{DH}\}$, as shown in Fig. 5.14. A circular path, which consists of all the boundary electrodes, connects all these devices.

The size of the electrode array derived by baseline algorithm is 20×15, and the number of electrodes required for this design is 68. The execution time of the complete bioassay is $(80 + 115t_m)$ seconds.

Compared with the baseline algorithm, the proposed design method can thus reduce the chip area by 59.7 %, the number of electrodes by 35.3 %, and the execution time by 57.8 % (when the value of t_m is set as 0.1).

Next, we study a protocol which represents a real-life PCR mixing ratio [34]. This bioassay is referred to as Bioassay 2. The mixing ratio of the eight components is written as [34]:

{Reaction buffer: Mag_Sulf: dNTPs: Forward primer: Reverse Primer: Optimase: Water: DNA} = {10 % : 10 % : 8 % : 0.8 % : 0.8 % : 1 % : 68.4 % : 1 %}. This ratio can be approximately as {51 : 51 : 41 : 4 : 4 : 5 : 351 : 5} on 512-scale; the corresponding mixing tree can be derived using the ratioed+ mixing algorithm (RMA) [27]. The simulation results of the PCR biochip designed for Bioassay 2 are shown in Table 5.2.

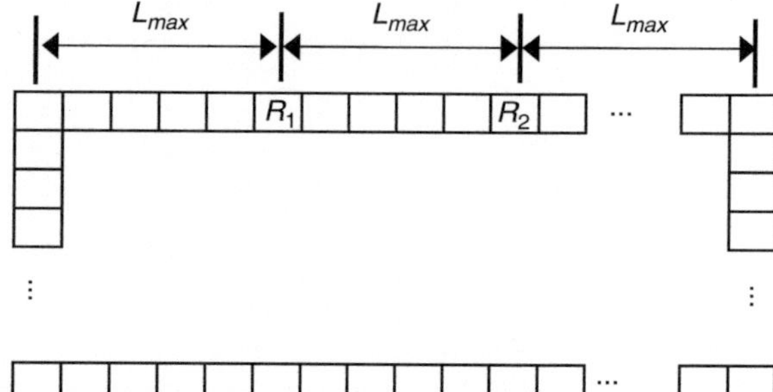

Fig. 5.14 The result of device placement derived by the baseline algorithm. The output ports of reservoir $1 \sim 7$, DE $1 \sim 2$ and the heater are placed on the boundary of the layout one by one. R_1 and R_2 here represent the output ports of two reservoirs

Table 5.2 Comparison of the PCR biochips derived by proposed method and the baseline algorithm

		Proposed method			Baseline algorithm		
Bioassay	Mixing ratio	Size of the biochip	No. electrodes	Execution time (s)	Size of the biochip	No. electrodes	Execution time (s)
1	$2:3:5:7:11:13:87$ [27]	11×11	44	$35 + 36t_m$ (38.6)	20×15	68	$80 + 115t_m$ (91.5)
2	$51:51:41:4:4:5:351:5$ [34]	13×11	49	$80 + 79t_m$ (87.9)	20×15	68	$105 + 163t_m$ (121.3)
3	$7:14:11$ [35]	11×6	30	$25 + 12t_m$ (26.2)	10×10	38	$25 + 37t_m$ (28.7)

*Numbers in parenthesis indicate execution time for $t_m = 0.1$

We consider another mixing protocol called Bioassay 3 [35], which has three input reagents/samples. The results are shown in Table 5.2. In all the three cases, the layout size, electrode count, and execution time of bioassays are significantly improved compared to the baseline method.

Since these three laboratory bioassays have relatively low numbers of devices that must be placed, the experimental results of the proposed algorithm are optimized by Approach 1 introduced in Sect. 5.3.4.

We further create three benchmarks that have a relatively large number of reservoirs and apply Approach 1, Approach 2, and the baseline algorithm to generate the resulting placements of the devices. Table 5.3 provides the results of the simulation. It is important to notice that, when searching the optimal device-placement results for Benchmarks 2 and 3, Approach 1 is impractical due to it high computational complexity. In Approach 2, the parameters α and β in 5.8 are set as 1 and 10, respectively. From Table 5.3, we find that Approach 2 can reduce the area of the biochip by 64.1–72.9 % compared with the baseline algorithm.

Table 5.3 Comparison of the results derived from the proposed Approach 1, Approach 2, and the baseline algorithm

Benchmark	(N_r, N_d, N_h)	Approach 1		Approach 2		Baseline algorithm	
		CPU time (s)	Size of the biochip	CPU time (s)	Size of the biochip	CPU time (s)	Size of the biochip
1	(10, 2, 1)	3,720	16×11	832	16×11	~ 0	25×25
2	(18, 2, 1)	NA	Computationally impractical	5,797	19×17	~ 0	30×30
3	(23, 2, 1)	NA	Computationally impractical	11,147	20×19	~ 0	35×40

$*N_r$, N_d, and N_h represent the number of reservoirs, DEs, and heaters, respectively

5.6.3 Defect Tolerance of Layouts for PCR Biochips

Because of manufacturing imperfections and degradation of electrodes, physical defects may occur on DMFBs. These defects can be classified into two categories based on their locations on the biochip. If a defect disconnects a droplet path into two "isolated parts" (i.e., droplets cannot be moved from one part to the other part) or the defect overlaps with a reservoir/DE/heater on the biochip, then it is defined as a "catastrophic defect". All other defects are defined as "non-catastrophic defects".

If a catastrophic defect occurs, the PCR biochip cannot be used further. On the layout designed for Bioassay 1, the positions of electrodes where defects are catastrophic are shown in Fig. 5.15a. For each of these three PCR biochips designed for Bioassays $1 \sim 3$, the total number of electrodes where defects are catastrophic is shown in Table 5.4.

If a non-catastrophic defect occurs, we can re-route the droplets and send them to their destinations, as shown in Fig. 5.15b. It is important to note that re-routing the droplet may increase the length of the transportation path, and reduce the degree of parallelism of fluid-handling operations. Therefore, compared to the defect-free biochip, the execution time of the bioassay on a biochip with a non-catastrophic defect is higher.

For each PCR biochip, we randomly insert a non-catastrophic defect into the biochip, then calculate the execution time of the bioassay. The defect-insertion simulations are executed for all the possible non-catastrophic defects on the layout. The average value and standard deviation for the execution time of running bioassays on biochips with defects are shown in Table 5.4. The value of t_m is set as 0.1. From the table, we find that the percentage of electrodes where defects are catastrophic is in the range from 33.3 % to 52.3 %. With the presence of a single non-catastrophic defect, the PCR biochip can be used via graceful degradation with 27.2~67.6 % average increase in the execution time of bioassay.

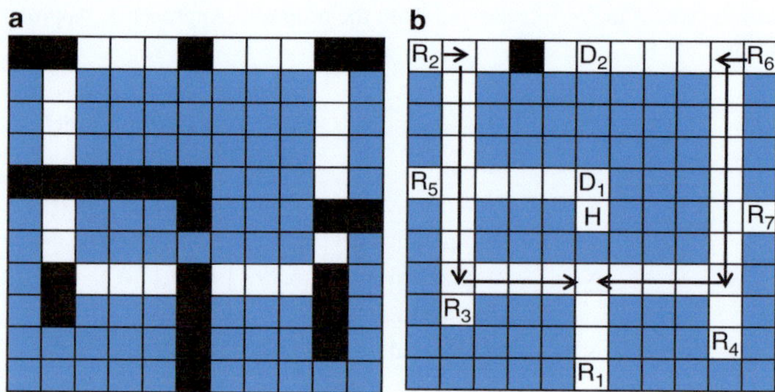

Fig. 5.15 (**a**) The positions of "catastrophic defects". If the electrodes at these positions have defects, the biochip cannot be used any more; (**b**) non-catastrophic defect can be by-passed by adjusting the routing of droplets

Table 5.4 Defect tolerance of PCR biochips

Bioassay	Total number of electrodes where defects are catastrophic	Standard deviation of Average execution time when a non-catastrophic defect is injected (s)	execution time when a non-catastrophic defect is injected (s)
1	23	64.7	7.9
2	23	111.8	14.4
3	10	36.1	6.5

5.7   Conclusion

This chapter has shown how cyberphysical integration of a PCR protocol on digital microfluidics can be used to execute on-chip bioassays reliably, despite the uncertainties inherent in fluidic operations such as dispensing operation and thermal cycling. The proposed design approach facilitates dynamic on-line decision making for the termination of thermal cycles in response to feedback from sensors. We have also presented new algorithms for device placement and layout design that safeguard the array against undesirable device interferences, avoid conflicts in droplet routing, and reduce bioassay execution time. The visibility of droplets in the monitoring system is also considered during the scheduling of fluid-handling operations. Simulation results on laboratorial protocols demonstrate that with minimized chip size and electrode count, the proposed design method can achieve high reliability and defect-tolerance.

References

1. Y. Luo, B. Bhattacharya, T.-Y. Ho and K. Chakrabarty, "Optimization of polymerase chain reaction on a cyberphysical digital microfluidic biochip", *Proc. IEEE/ACM International Conference on Computer-Aided Design*, pp. 622–629, 2013.
2. J. Lage, J. Leamon, T. Pejovic, S. Hamann, M. Lacey, D. Dillon, et. al, "Whole genome analysis of genetic alterations in small DNA samples using hyperbranched strand displacement amplification and array-GH", *Genome Res.* Issue 13, pp. 294–307, 2003.
3. J. Berthier, *Micro-Drops and Digital Microfluidics*, Norwich, NY: William Andrew, 2008.
4. I. Erill, S. Campoy, J. Rus, L. Fonseca, A. Ivorra, Z. Navarro, J. Plaza, J. Aguilo, and J. Barbe, "Development of a CMOS-compatible PCR chip: comparison of design and system strategies", *Journal of Micromechanics and Microengineering*, Volume 14, Number 11, pp. 1–11, 2014.
5. C. Zhang and D. Xing, "Miniaturized PCR chips for nucleic acid amplification and analysis: latest advances and future trends", *Nucleic Acids Research*, Vol. 35, No. 13, pp. 4223–4237, 2007.
6. D. Brassard, L. Malic, C. Miville-Godin, F. Normandin, and T. Veres, "Advanced EWOD-based digital microfluidic system for multiplexed analysis of biomolecular interactions", *IEEE International Conference on Micro Electro Mechanical Systems (MEMS)*, pp. 153–156, 2011.
7. D. Jary, A. Chollat-Namy, Y. Fouillet, J. Boutet, C. Chabrol, G. Castellan, D. Gasparutto, and C. Peponnet, "DNA repair enzyme analysis on EWOD fluidic microprocessor", *Proceedings of the NSTI-Nanotech Conference*, vol. 2, pp. 554–557, 2006.
8. K. Chakrabarty and F. Su, *Digital Microfluidic Biochips: Synthesis, Testing, and Reconfiguration Techniques*, Boca Raton, FL : CRC Press, 2006.
9. Y. Luo, K. Chakrabarty, and T.-Y. Ho, "Dictionary-based error recovery in cyberphysical digital-microfluidic biochips", *Proc. IEEE/ACM International Conference on Computer-Aided Design*, pp. 369–376, 2012.
10. R. Liu, J. Yang, R. Lenigk, J. Bonanno, and P. Grodzinski, "Self-contained, fully integrated biochip for sample preparation, polymerase chain reaction amplification, and DNA microarray detection", *Anal. Chem.*, Issue 76, pp. 1824–1831, 2004.
11. L. Malic, T. Veres, and M. Tabrizian, "Detection of DNA hybridization on a configurable digital microfluidic biochip using SPR imaging", *International Conference on Miniaturized Systems for Chemistry and Life Sciences*, pp. 829–831, 2008.
12. L. Malic, T. Veres, and M. Tabrizian, "Two-dimensional droplet-based surface plasmon resonance imaging using electrowetting-on-dielectric microfluidics", *Lab on a Chip*, Issue 9, pp. 473–475, 2009.
13. K. Choi, A. Ng, R. Fobel, and A. Wheeler, "Digital microfluidics", *Annual Review of Analytical Chemistry*, Vol. 5, pp. 413–440, 2012.
14. M. Shamsi, K. Choi, A. Ng, A. Wheeler, "A digital microfluidic electrochemical immunoassay", *Lab on a Chip*, Issue 14, pp. 547–554, 2014.
15. S. Koster, F. Angile, H. Duan, J. Agresti, A. Wintner, C. Schmitz, A. Rowat, C. Merten, D. Pisignano, A. Griffiths. and D. Weitz, "Drop-based microfluidic devices for encapsulation of single cells", *Lab on a Chip*, vol. 8, pp. 1110–1115, 2008.
16. R. Daniel, M. Dines, and H. Petach, "The denaturation and degradation of stable enzymes at high temperatures", *Biochem J.*, vol. 317, Issue 1, pp. 1–11, 1996.
17. F. Ji, M. Juntunen, and I. Hietanen, "Evaluation of electrical crosstalk in high-density photodiode arrays for X-ray imaging applications", *Nuclear Instruments and Methods in Physics Research*, volume 610, issue 1, pp. 28–30, 2009.
18. R. Evans et al., "Optical detection heterogeneously integrated with a coplanar digital microfluidic lab-on-a-chip platform", *Proc. IEEE Sensors Conf.*, pp. 423–426, Oct. 2007.
19. S. Koester, L. Schares, C. Schow, G. Dehlinger, and R. John, "Temperature-dependent analysis of Ge-on-SOI photodetectors and receivers", *IEEE International Conference on Group IV Photonics*, pp. 179–181, 2006.

20. C. Zhang and D. Xing, "Single-molecule DNA amplification and analysis using microfluidics", *Chem Rev.*, Issue. 110, vol. 8, pp. 4910–4947, 2010.
21. D. Woide, A. Zink, and S. Thalhammer, "Technical note: PCR analysis of minimum target amount of ancient DNA", *Am J Phys Anthropol*, volume 142, Issue 2, pp. 321–327, 2010.
22. J. Webster, M. Burns, D. Burke, and C. Mastrangelo, "Monolithic capillary electrophoresis device with integrated fluorescence detector", Anal Chem., volume 73, Issue 7, pp. 1622–1626, 2001.
23. U.-C. Yi and C.-J. Kim, "Soft printing of droplets pre-metered by electrowetting", *Sensors and Actuators A: Physical*, Volume 114, Issues 2–3, pp. 347–354, 2004.
24. I. Erill, S. Campoy, J. Rus, L. Fonseca, A. Ivorra, Z. Navarro, J. Plaza, J. Aguilo, and J. Barbe, "Development of a CMOS-compatible PCR chip: comparison of design and system strategies", *Journal of Micromechanics and Microengineering*, Volume 14, Number 11, pp. 1–11, 2004.
25. M. Garey and D. Johnson, *Computers and Intractability: A Guide to the Theory of NP-Completeness*. W.H. Freeman & Company, 1979.
26. P. Bhowmick, A. Biswas, B. Bhattacharya, "ICE: The isothetic convex envelope of a digital object", *International Conference on Computing: Theory and Applications*, pp. 219–223, 2007.
27. S. Roy, B. Bhattacharya, P. Chakrabarti, and K. Chakrabarty, "Layout-aware solution preparation for biochemical analysis on a digital microfluidic biochip", *Proc. IEEE International Conference on VLSI Design*, pp. 171–176, 2011.
28. Y.-L. Hsieh, T.-Y. Ho, and K. Chakrabarty, "Design Methodology for Sample Preparation on Digital Microfluidic Biochips", *IEEE International Conference on Computer Design*, pp. 189–194, 2012.
29. Y. Luo, K. Chakrabarty, and T.-Y. Ho, "Design of cyberphysical digital-microfluidic biochips under completion-time uncertainties in fluidic operations", pp. 44–50, *Proc. IEEE/ACM Design Automation Conference*, 2013.
30. P. Paik, V. Pamula, and R. Fair, "Rapid droplet mixers for digital microfluidic systems", *Lab on a Chip*, vol. 3, pp. 253–259, 2003.
31. P. Paik, V. Pamula, M. Pollack, and R. Fair, "Electrowetting-based droplet mixers for microfluidic systems", *Lab on a Chip*, vol. 3, Issue 1, pp. 28–33, 2003.
32. E. Bolton, G. Sayler, D. Nivens, J. Rochelle, S. Ripp, and M. Simpson, "Integrated CMOS photodetectors and signal processing for very low-level chemical sensing with the bioluminescent bioreporter integrated circuit", *Sens Actuators B Chem*, vol. 85, Issue 1, pp. 179–185, 2002.
33. P. Paik, V. Pamula and R. Fair, "Rapid droplet mixers for digital microfluidic systems", *Lab on a Chip*, vol. 3, pp. 253–259, 2003.
34. (Optimase Master Mix Calculator) http://www.mutationdiscovery.com/md/MD.com\discretionary-/screens/optimase/MasterMixCalculator.jsp?action=none
35. Y.-L. Hsieh, T.-Y. Ho and K. Chakrabarty, "A reagent-saving mixing algorithm for preparing multiple-target biochemical samples using digital microfluidics", *IEEE Transactions on Computer-Aided Design of Integrated Circuits and Systems*, vol. 31, pp. 1656–1669, 2012.

Chapter 6
Pin-Count Minimization
for Application-Independent Chips

In this chapter, we propose design methods for pin-limited general-purpose microfluidic biochips. The number of control pins used to drive electrodes is a major contributor to fabrication cost for disposable biochips in a highly cost-sensitive market. Most prior work on pin-limited biochip design determines the mapping of a small number of control pins to a larger number of electrodes according to the specific schedule of fluid-handling operations and routing paths of droplets. Such designs are therefore specific to the bioassay application, hence sacrificing some of the flexibility associated with digital microfluidics. We propose a design method to generate an application-independent pin-assignment configuration with a minimum number of control pins. Layouts of commercial biochips and laboratory prototypes are used as case studies to evaluate the proposed design method for determining a suitable pin-assignment configuration. Compared with previous pin-assignment algorithms, the proposed method can reduce the number of control pins and facilitate the "general-purpose" use of digital microfluidic biochips for a wider range of applications.

6.1   Motivation and Related Prior Work

In recent years, the complexity of digital microfluidic biochips continues to increase as new applications are targeted by this platform [1, 2]. For example, recently announced commercial products contain up to 100,000 electrodes [3]. In order to ensure complete reconfigurability and the ability to run any given bioassay on the digital microfluidic platform (i.e., "general-purpose use"), it is desirable that every electrode be controlled by an independent pin. However, a one-to-one mapping between control pins and electrodes (referred to as direct-addressing pin-assignment) is not practical for low-cost disposable biochips. A large number of control pins leads to high fabrication cost, and interconnect routing problems [4].

© Springer International Publishing Switzerland 2015
Y. Luo et al., *Hardware/Software Co-Design and Optimization for Cyberphysical Integration in Digital Microfluidic Biochips*, DOI 10.1007/978-3-319-09006-1_6

In order to reduce the number of control pins and to control the digital microfluidic array without significantly affecting the reconfigurability of the biochip, a number of design of optimization techniques have been published in the literature. These techniques can be categorized as being either bioassay-independent [5, 6] or bioassay-specific [7–9]. In bioassay-independent techniques such as the use of a bus-phase addressing [5] or cross-referencing [6], the number of control pins required for addressing the electrodes is independent of the target application. For example, analogous to row/column-based addressing in memories, cross-referencing requires $m + n$ pins for an $m \times n$ array of electrodes. Bioassay-specific pin-assignment methods lead to fewer control pins since they utilize knowledge about the operation schedule, module placement, and droplet routing pathways of the target bioassay.

Prior methods on bioassay-specific pin assignment suffer from three main drawbacks. First, these techniques are not effective for the design of multi-functional biochips, which can be reconfigured post-fabrication for different applications by loading the appropriate control software. General purpose (application-independent) biochips, where software can be used as a differentiator, offer the promise of higher production volume and reduced cost. Second, fluid-handling operations on an application-specific biochip are constrained by the pre-determined pin-assignment. Hence it is impossible to perform post-fabrication tuning of the bioassay protocol, schedule, and droplet routing. Finally, it is difficult to estimate the number of control pins a priori since the number of pins is application-dependent. The cross-referencing technique described above is application-independent; however, it requires a special electrode structure which both top and bottom plates are divided into discrete electrode arrays. This results in increased complexity and higher manufacturing cost [6].

To overcome the above drawbacks, we propose a new method to generate pin-assignment configurations. This method does not depend on actuation sequences of electrodes, or does the scheduling and the routing of droplets. Any target application can be mapped to the array without any restriction on droplet manipulation. Therefore, the degree of freedom for droplet movement is maximized.

The main contributions of this chapter are as follows:

1. An analysis of pin-actuation conflicts and freedom of movement of a single droplet in all feasible directions (Sect. 6.2).
2. The derivation of necessary and sufficient conditions for control-pin sharing to ensure high flexibility in the concurrent movement of two droplets (Sect. 6.2).
3. An integer linear programming model for designing a pin-assignment with the smallest number of pins (Sect. 6.3).
4. A graph-theoretic method to formulate an acceptance test for a pin-assignment configuration and a lower bound on the number of pins (Sect. 6.4).
5. A heuristic algorithm that generates a pin-assignment configuration for biochips (Sect. 6.4).
6. Extension of the study from $1\times$ volume droplets to $2\times$ and even larger droplets (Sect. 6.5).

7. A scheduling algorithm that can be applied to biochips with pin-constraints (Sect. 6.6).
8. Results for commercial biochips and experimental prototypes (Sect. 6.7).

6.2 Analysis of Pin-Assignment

In this section, we first discuss the relationship between droplet movement and voltages applied to the electrodes. Next, the concept of pin-actuation conflict is introduced. Finally, in order to determine the conditions that guarantee conflict-free pin-assignment, several pin-assignment configurations are analyzed.

6.2.1 Pin-Actuation Conflicts

Appropriate control voltages must be applied to a group of electrodes to manipulate a droplet that currently resides on an electrode $\mathscr{E}$. According to the working principle of electrowetting-on-dielectric (EWOD)-based microfluidic biochips, movements of a droplet are determined by the group of electrodes that are directly in contact with the droplet. Suppose a droplet of unit volume (referred to as a "1×" droplet) is held on electrode $\mathscr{E}$. Then the electrode group that can determine the movement of the 1× droplet consists of the central electrode $\mathscr{E}$ and all its non-diagonally adjacent electrodes. Each non-diagonally adjacent electrode is a possible destination for the droplet. An example is presented in Fig. 6.1a. Each square in Fig. 6.1a presents an electrode on the microfluidic biochip; letters (such as "A", "B" and "C") stand for the names of control pins that are connected to the corresponding electrodes. The control voltages applied to the control pins are either "High", "Low" or "don't-care". Here we introduce two definitions:

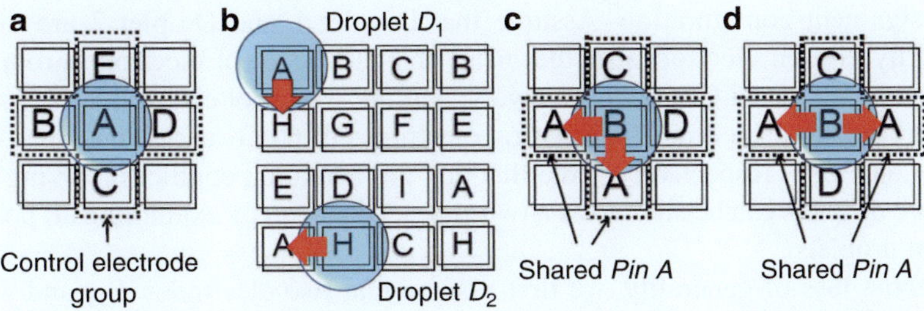

Fig. 6.1 (**a**) A central electrode and its non-diagonally adjacent electrodes comprising an electrode group; (**b**) an example of pin-actuation conflicts; (**c**) an example where two diagonally adjacent electrodes in the same CEG share one pin (Pin A); (**d**) an example where two non-adjacent electrodes in the same CEG share one pin (Pin A)

Control electrode group (CEG): An electrode on which the droplet rests at any given time and all other electrodes that have direct contact with the droplet are defined as the elements of the control electrode group (CEG).

For a $1\times$ droplet, its CEG includes the central electrode on which the droplet rests and all the non-diagonally adjacent electrodes, see Fig. 6.1a. Most bioassays executed on biochips only include the transportation of $1\times$ droplets, hence we only discuss the case of $1\times$ droplets on the biochip in the following sections (Sect. 6.2.2 through Sect. 6.4). The pin-assignment problem associated with biochips that have larger droplets on them is discussed in Sect. 6.5.

Control pin group (CPG): All pins that are connected to the electrodes in the CEG are defined as the elements of the control pin group (CPG).

When multiple fluid-handling operations are performed on a biochip with a given pin-assignment configuration, pin-actuation conflicts must be considered. An example is shown in Fig. 6.1b, where droplets D_1 and D_2 are on the array, and they are scheduled to move concurrently in the directions of the arrows. The groups of control pins for D_1 and D_2 are {A, B, H} and {A, C, D, H}, respectively. Note that Pin A and Pin H are common (shared) control pins for droplets D_1 and D_2. In order to move D_1 in the designated direction, the voltages applied to A, B, H should be set as "Low", "Low", and "High", respectively. Similarly, in order to move droplet D_2, the voltages of A, C, D, H should be set as "High", "Low", "Low", and "Low", respectively. Thus, the movement of D_1 requires the application of "High" voltage on Pin H, while the movement of D_2 requires the application of "Low" voltage on Pin H. The movements of D_1 and D_2 cannot be implemented concurrently since it is impossible to apply these different voltages to Pin H at the same time. This problem is referred to as "conflict in pin-actuation signals".

6.2.2 Control-Pin Sharing and Concurrent Movement of Droplets

To analyze pin-actuation conflicts, we consider an electrode array with any arbitrary pin-assignment configuration. Assume that Droplet 1 and Droplet 2 are on two arbitrarily chosen electrodes in the array and their control electrode groups are written as CEG_1 and CEG_2, respectively, and these control electrode groups have no overlap with each other. The central electrodes of CEG_1 and CEG_2 are written as E_{1C} and E_{2C}, respectively. Note that any pin-actuation conflict involving more than two droplets can be studied as a two-droplet problem by examining all possible pairs of droplets.

Without loss of generality, we first assume that two electrodes E_{11} and E_{12} in CEG_1 share one pin (Pin A). Then we have the following scenarios, which are exhaustively enumerated by Cases (1.a)~(1.c):

(1.a) E_{11} and E_{12} are non-diagonally adjacent electrodes: In this case, the movement of a droplet along some directions cannot be achieved due to the electric shorting of adjacent electrodes [10].

(1.b) E_{11} and E_{12} are diagonally adjacent electrodes: An example can be found in Fig. 6.1c. Two non-diagonally adjacent electrodes of the droplet are controlled by Pin A. If "High" signal is applied to Pin A, the two electrodes that are connected to Pin A will pull the droplet from two directions at the same time. The droplet may undergo unwanted and unpredictable movement.

(1.c) E_{11} and E_{12} are non-adjacent electrodes: An example is shown in Fig. 6.1d. Two non-adjacent electrodes of the droplet are controlled by Pin A. The droplet may be split if Pin A activates the two electrodes at the same time.

According to the above discussion for Cases (1.a)∼(1.c), we conclude that in order to avoid unwanted movement or splitting of droplets and also guarantee the flexibility of droplet movements, electrodes in the same CEG cannot share control pins.

Next we assume that CEG_1 and CEG_2 share one pin (without loss of generality, we assume that the shared pin is Pin A). Then we have the following scenarios, exhaustively enumerated by Cases (2.a)∼(2.c):

(2.a) Pin A is connected to both E_{1C} and E_{2C}: An example can be found in Fig. 6.2a. The two droplets can be moved concurrently to any non-diagonally adjacent electrode without pin-actuation conflicts. Thus, there are 16 possible concurrent movements of the pair of droplets, and no unwanted movement or splitting will occur for these 16 concurrent movements.

(2.b) Pin A is neither connected to E_{1C} nor E_{2C}: An example can be found in Fig. 6.2b. Suppose Droplet 1 and Droplet 2 are scheduled to move in the directions indicated by the arrows. For the movement of Droplet 1, the control voltages applied to Pins A, B, C, D, and E should be set as "High", "Low", "Low", "Low", and "Low", respectively. Similarly, for the movement of Droplet 2, the control voltages on Pins A, F, G, H, and I should be set as "Low", "Low", "Low", "High", and "Low", respectively. Thus the status of Pin A corresponding to the movements of Droplet 1 and Droplet 2 are different. In order to avoid a conflict, Droplet 1 and Droplet 2 must be moved in different clock cycles. Assume that Droplet 1 is moved first; a "High" voltage will be applied to Pin A. To keep Droplet 2 stay at its current position, a "High" voltage will also be applied to the electrode under Droplet 2 (i.e., a "High" voltage will be applied to Pin F). Note that when a high voltage is applied to the electrode under the droplet, the droplet will remain at its current position without movement or splitting, even if high voltage is applied to one of its adjacent electrodes; this scenario has been experimentally demonstrated [11]. After Droplet 1 has arrived at its destination electrode, the movement operation for Droplet 2 will be performed. Thus, the number of all possible concurrent movements for the droplet pair is $1 + 3 \times 3 = 10$, and no unwanted movement or splitting will occur for these 10 concurrent movements.

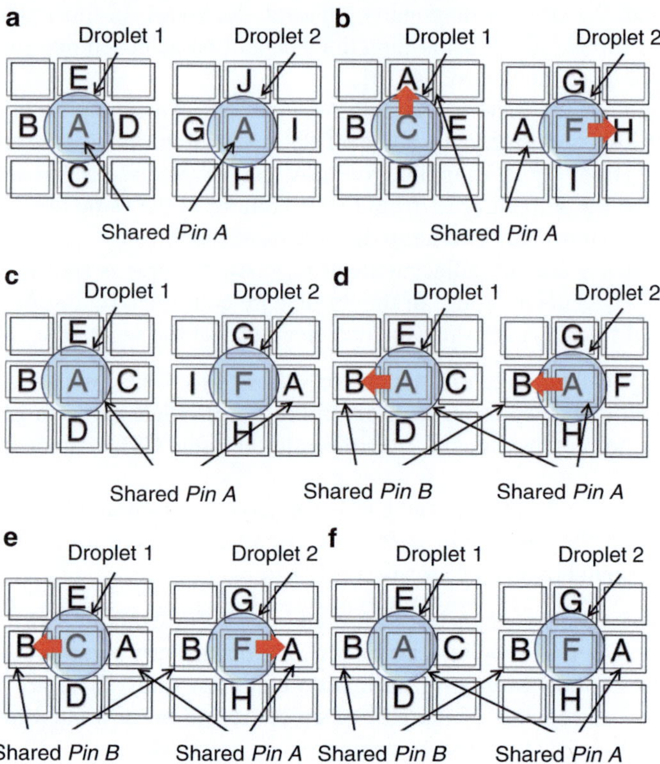

Fig. 6.2 Six pin-assignment configurations that are analyzed in Cases (2.a)∼(2.c) and Cases (3.a)∼(3.e): (**a**) Case (2.a); (**b**) Case (2.b); (**c**) Case (2.c); (**d**) Case (3.a); (**e**) Case (3.c); and (**f**) Case (3.d)

(2.c) Pin A is connected to E_{1C} or E_{2C}: Without loss of generality, we assume that Pin A is connected to E_{1C}. An example is shown in Fig. 6.2c. Droplet 1 can be freely moved in any non-diagonal directions, while Droplet 2 can be moved to the electrodes that are controlled by Pin G, Pin H, and Pin I. Thus, for the droplet pair, the number of all possible concurrent movements is = 12. No unwanted movement or splitting will occur for these 12 concurrent movements.

Based on the above analysis for Cases (2.a)∼(2.c), we find that when two CEGs share one control pin, and electrodes in the same CEG do not share any control pin, unwanted movement or splitting of droplet will not occur. The number of all possible concurrent movements for the droplet pair varies from 10 to 16. Even though in Cases (2.b) and (2.c) the flexibilities of droplet movements are not as high as direct-addressing biochips, they are still adequate for the concurrent manipulation of droplets.

Finally, we assume that CEG_1 and CEG_2 share two pins (Pin A and Pin B). Then we have the following scenarios, exhaustively enumerated by Cases (3.a)∼(3.e):

(3.a) E_{1C} and E_{2C} both are connected to Pin A (or Pin B): Without loss of generality, we assume that both E_{1C} and E_{2C} are connected to Pin A. One example can be found in Fig. 6.2d. When Droplet 1 is moved in the direction indicated by the arrow, the control voltages applied to Pins A, B, C, D, and E should be set as "Low", "High", "Low", "Low", and "Low", respectively. In this case, the electrode under Droplet 2 is applied "Low" voltage, while one of the non-diagonal electrodes (which is connected to Pin B) is applied "High" voltage. Therefore, no matter where the scheduled movement direction points, Droplet 2 has to be moved in the direction indicated by the arrow. In this case, the unwanted movement of the droplet may occur. The number of all possible concurrent movements for the droplet pair is $1 + 3 \times 3 = 10$. To avoid unwanted movement of droplets that may be located on any two arbitrary positions of the layout, the pin-assignment configurations discussed in this case should be forbidden.

(3.b) Pin A connects to E_{1C} and Pin B connects to E_{2C} (or vice versa): Assume Pin A is also connected to electrode $E_{2L} \in \mathrm{CEG}_2$, and Pin B is also connected to electrode $E_{1L} \in \mathrm{CEG}_1$. Then CEG_{2L} denotes the control electrode group whose central electrode is E_{2L}. Since E_{2C} and E_{2L} are two adjacent electrodes, we find that $E_{2C} \in \mathrm{CEG}_{2L}$. Therefore, CEG_1 and CEG_{2L} share two pins and their central electrodes are both controlled by Pin A. The pin-assignment configuration for CEG_1 and CEG_{2L} is the same as the configuration discussed in Case (3.a), and may lead to the unwanted movement of a droplet. The pin-assignment configurations discussed in this case should also be forbidden.

(3.c) Pin A and B are connected to neither E_{1C} nor E_{2C}: An example is shown in Fig. 6.2e. Assume that Droplet 1 and Droplet 2 are scheduled to move in the directions indicated by the arrows. For the movement of Droplet 1, the control voltages applied to Pins A, B, C, D, and E should be set as "Low", "High", "Low", "Low", and "Low", respectively. Similarly, for the movement of Droplet 2, the control voltages on Pins A, B, F, G, and H should be set as "High", "Low", "Low", "Low", and "Low", respectively. Therefore, the status of Pin A and Pin B corresponding to the movements of Droplet 1 and Droplet 2 are different. To avoid a conflict, Droplet 1 and Droplet 2 must be moved in different clock cycles. Assume that Droplet 1 is moved first; a "High" voltage will be applied to Pin B. In order to keep Droplet 2 stay at its current position, a "High" voltage will also be applied to the electrode under Droplet 2. Therefore, for the pin-assignment configuration discussed in this case, one droplet can be moved freely, while the other droplet may have to stall on the current electrode. The number of all possible concurrent movements for the droplet pair is $1 + 1 + 2 \times 2 = 6$. Note that even though droplets will not undergo unwanted movement or splitting in this case, the number of possible movements is relatively low. This shows that the flexibility of droplet movement in this case is rather limited, hence we should avoid such a pin-assignment configuration.

(3.d) Pin A is connected to E_{1C} or E_{2C}, and Pin B is neither connected to E_{1C} nor E_{2C}: Without loss of generality, we assume that Pin A is connected to E_{1C}. Figure 6.2f shows an example. In this pin-assignment configuration, unwanted movement or splitting of droplets will not occur. The number of possible concurrent movements in this case is $1 + 6 = 7$, as explained below. Droplet 1 and Droplet 2 can be moved to the electrodes controlled by Pin B concurrently (1 possible concurrent movement); Droplet 1 can be moved to the electrodes that are controlled by Pins E, C, and D; while Droplet 2 can be moved to the electrodes controlled by Pins G and H (3×2 possible concurrent movements).

(3.e) Pin B is connected to E_{1C} or E_{2C}, and Pin A is neither connected to E_{1C} nor E_{2C}: This case can be analyzed in the same way as Case (3.d). The number of all possible concurrent movements for the droplet pair is 7.

Based on the above analysis, we conclude that when two CEGs share two pins, for Case (3.a) and Case (3.b), droplets may undergo unwanted movement. For Cases (3.c)~(3.e), unwanted movement or splitting of droplet will not occur. However, the number for possible concurrent movements is no more than 7. Comparing with a direct-addressing biochip, the flexibilities of droplet movements in Cases (3.c)~(3.e) are relatively low. Hence such pin assignments are not considered in the biochip designs studied in this chapter.

Based on the above discussion, we obtain the following lemma, which provides a necessary and sufficient condition for the acceptance of any arbitrary pin-assignment configuration with two droplets on two arbitrary positions of the biochip. The goal is to guarantee significant flexibility for the concurrent movement of droplets, and also prevent unwanted splitting or movement of droplets.

Lemma 6.1. *Consider an electrode array with any arbitrary pin-assignment configuration. Suppose two droplets are located at any two arbitrarily chosen electrodes. The following constraints are necessary and sufficient to avoid any unwanted splitting or movement of droplets, to permit the movement of each droplet along any feasible direction in the array, and to guarantee that the number of possible concurrent movements is no less than 10:*

Constraint 1. In the same control electrode group, any two electrodes cannot be connected to the same control pin.

Constraint 2. Any two non-overlapping electrode groups cannot share more than one pin.

Proof. The necessity of Constraint 1 can be proven by *reductio ad absurdum*. We assume that two electrodes in the same CEG share one pin. Suppose unwanted splitting and movement of droplets can be avoided, and each droplet can be moved along any feasible direction in the array. From the discussion of Cases (1.a)~(1.c), we can find that droplets may undergo unwanted splitting or movement, or the movement of a droplet along some directions can never be achieved. Hence we have reached a contradiction.   □

The necessity of Constraint 2 also can be proven by *reductio ad absurdum*. We assume that two CEGs share k ($k > 1$) control pins. From the discussion of Cases (3.a)~(3.e), we find that in some cases, droplets may undergo unwanted splitting or movement. In every other case, the number of possible concurrent movements for the two droplets is no more than 7. The sufficiency of Constraint 1 and Constraint 2 is proven based on the analysis for Cases (2.a)~(2.c). When Constraint 1 and Constraint 2 are satisfied, unwanted movement or splitting of droplets will not occur. The number of all possible concurrent movements for the droplet pair is at least 10, and each droplet can be moved along any feasible direction in the array.

Based on the above discussion, the necessity and sufficiency of Constraint 1 and Constraint 2 in Lemma 6.1 are proven.

Note that Constraint 1 and Constraint 2 also ensure that when one droplet is split, the other droplet does not undergo any unwanted splitting or movement. Since the movement of a droplet from one electrode to its adjacent electrode is the basic operation for mixing, dispensing, and transportation operations, Constraint 1 and Constraint 2 can ensure the maximum degree of freedom for any concurrent fluid-handing operations involving any two droplets.

6.3   ILP Model for Pin-Assignment

In this section, we develop an integer linear programming (ILP) model to optimally solve the optimization problem for pin-assignment. As explained later, the model forms the basis for evaluating heuristic solutions. On an $M \times N$ electrode array, let $x_{i,m,n}$ be a binary variable defined as below.

$$x_{i,m,n} = \begin{cases} 1, & \text{if Pin } i \text{ is connected to the electrode at the} \\ & m^{\text{th}} \text{ row and } n^{\text{th}} \text{ column of the array} \\ 0, & \text{otherwise} \end{cases}$$

where $1 \leq i \leq L$. The parameter L is the maximum possible index for the number of control pins. The value of L can be set to an easily-determined loose upper bound (e.g., $M \times N$).

The index of the control pin connected to the electrode at the mth row and n^{th} column of the array is defined as $P_{m,n}$. It can be expressed as $P_{m,n} = \sum_{i=1}^{L} i \cdot x_{i,m,n}$. The total number of pins that are assigned to electrodes is equal to the maximum value of $P_{m,n}$ (where $1 \leq m \leq M$ and $1 \leq n \leq N$). Hence, the total number of pins assigned to electrodes in the layout (i.e., N_{tol}) can be written as: $N_{\text{tol}} = \text{Max}_{1 \leq i \leq L, 1 \leq m \leq M, 1 \leq n \leq N} \{i \cdot x_{i,m,n}\}$. Since the target of the ILP model is to derive a feasible pin-assignment configuration that has the minimum number of control pins. For a pin-limited digital microfluidic biochip, the objective function of the ILP model is defined as:

$$\text{minimize: } N_{\text{tol}} \tag{6.1}$$

Next we map Constraint 1 and Constraint 2 of Sect. 6.1 into inequalities of the ILP model. The electrode group whose central electrode is at the mth row and n^{th}

column of the array is written as $EG_{m,n}$. In any electrode group $EG_{m,n}$, the number of electrodes that are connected to any Pin i is defined as $N_{i,m,n}$, which can be written as:

$$N_{i,m,n} = x_{i,m,n} + x_{i,m+1,n} + x_{i,m-1,n} + x_{i,m,n+1} + x_{i,m,n-1}. \tag{6.2}$$

Therefore, Constraint 1 of Sect. 6.1 can be written as the following set of constraints: $N_{i,m,n} \leq 1, \forall 1 \leq i \leq L, 1 \leq m \leq M, 1 \leq n \leq N$.

The Constraint 2 of Sect. 6.1 can be further derived as follows. For two electrode groups $EG_{m,n}$ and $EG_{m+p,n+q}$, if p and q are integers and $|p|+|q| \geq 3$, then these two electrode group are non-overlapping. Based on the definition given by Equation 6.2, for $EG_{m+p,n+q}$, the number of electrodes that are connected to Pin i is $N_{i,m+p,n+q}$. For $EG_{m,n}$ and $EG_{m+p,n+q}$, the numbers of electrodes that are connected to another Pin j are written as $N_{j,m,n}$ and $N_{j,m+p,n+q}$, respectively. Assume that Constraint 2 in Sect. 6.1 is violated, and $EG_{m,n}$ and $EG_{m+p,n+q}$ share Pin i and Pin j. Then there will be: $N_{i,m,n} \geq 1, N_{i,m+p,n+q} \geq 1, N_{j,m,n} \geq 1$, and $N_{j,m+p,n+q} \geq 1$. Therefore, when Constraint 2 is violated, there must exist integers i, j, m, n, p, and q, such that $N_{i,m,n} + N_{i,m+p,n+q} + N_{j,m,n} + N_{j,m+p,n+q} \geq 4$. In order to make sure Constraint 2 is not violated, we have the following inequalities:

$$N_{i,m,n} + N_{i,m+p,n+q} + N_{j,m,n} + N_{j,m+p,n+q} \leq 3,$$
$$\forall 1 \leq i \neq j \leq L, 1 \leq m \leq M, 1 \leq n \leq N, |p| + |q| \geq 3. \tag{6.3}$$

Using standard techniques, the objective function given by (6.1) can be linearized. This completes the ILP model for optimization. For the $M \times N$ microfluidic array, the number of CEGs is MN. The number of inequalities derived from Constraint 1 and Constraint 2 are $O(MNL)$ and $O(M^2N^2L^2)$, respectively. Thus for the ILP model introduced above, the number of variables in the ILP model is $O(MNL)$, and the number of constraints is $O(M^2N^2L^2)$. It is important to note that, since L is a loose upper bound for the number of pins, we have $L = O(MN)$. The ILP model is clearly not scalable for a large problem instance; nevertheless, we use it to evaluate the quality of the heuristic solution that will be discussed in the next section.

6.4 Heuristic Optimization Method

The ILP model introduced in Sect. 6.3 has high computational complexity. The CPU time is unacceptable for large biochips. In this section, we discuss a heuristic algorithm to efficiently generate a pin-assignment configuration; the pseudocode for this algorithm is shown in Fig. 6.3. The algorithm includes two phases; the first phase is to establish a lower bound on the number of pins, and the second phase is to construct the pin-assignment configuration in a greedy fashion. In order

---

1: **Phase 1**: Derive the lower bound N_L on the number of control pins;

2: *PinAvailable* = {Pin1, Pin 2, ... Pin N_L}; // Initialize the set of pins that can be assigned to the layout

3: **Phase 2**: Construct a feasible pin-assignment configuration by greedy algorithm;

4: Give each electrode an index number, and name electrodes as E_1, E_2, ..., $E_{N_{\text{layout}}}$; // For example, we can number electrodes one by one from upper right corner of the layout to the lower left corner

5: ElectrodesNeedAssigned = {E_1, E_2, ..., $E_{N_{\text{layout}}}$}; // This is the set of electrodes that need to be assigned pins

6: ElectrodesAlreadyAssigned = $\emptyset$; // This is the set of electrodes that have already been assigned pins

7: Randomly select an electrode E_R and assign Pin 1 to it;

8: Set ElectrodesNeedAssigned = {E_R}; // E_R is the randomly selected "seed" of the pin-assignment process.

9: **while** ElectrodesNeedAssigned $\neq \emptyset$ **do**

10:     Select the electrode E_{neighbor} that has the minimum index from the neighboring electrodes of ElectrodesAlreadyAssigned;

11:     **for each** Pin i in *PinAvailable* **do**

12:         Assign Pin i to E_{neighbor};

13:         *CheckFeasible*; // Check whether the pin-assignment is a feasible solution

14:         **if** *CheckFeasible* = True **then**

15:             ElectrodesNeedAssigned = ElectrodesNeedAssigned$-E_{\text{neighbor}}$;

16:             ElectrodesAlreadyAssigned = ElectrodesAlreadyAssigned$+E_{\text{neighbor}}$;

17:             Break;

18:         **end if**

19:     **end for**

20:     **if** no pins in *PinAvailable* can be assigned to E_{neighbor} for a feasible solution **then**

21:         *PinAvailable* = *PinAvailable* + P_{new}; // Add a new pin to the layout

22:         Assign P_{new} to E_{neighbor}; //Assign the newly added pin to E_{neighbor}

23:         ElectrodesNeedAssigned = ElectrodesNeedAssigned $-E_{\text{neighbor}}$;

24:         ElectrodesAlreadyAssigned = ElectrodesAlreadyAssigned$+E_{\text{neighbor}}$;

25:     **end if**

26: **end while**

Fig. 6.3 Pseudocode for the construction of the pin-assignment configuration using a greedy algorithm

to improve the efficiency of the greedy algorithm, we propose a mapping from a pin-assignment configuration to an undirected graph, and then we use the graph to ascertain whether the pin-assignment configuration is a feasible solution. A lower bound on the number of pins from the graph model is also described.

For simplicity, two operators will be defined. First, the pin-assignment configuration is defined as the operator $\mathscr{C}$, which is a mapping from the set of all electrodes on the chip (E^*) to the set of pins that are assigned to the electrodes (P).

$$\mathscr{C} : E^* \rightarrow P \tag{6.4}$$

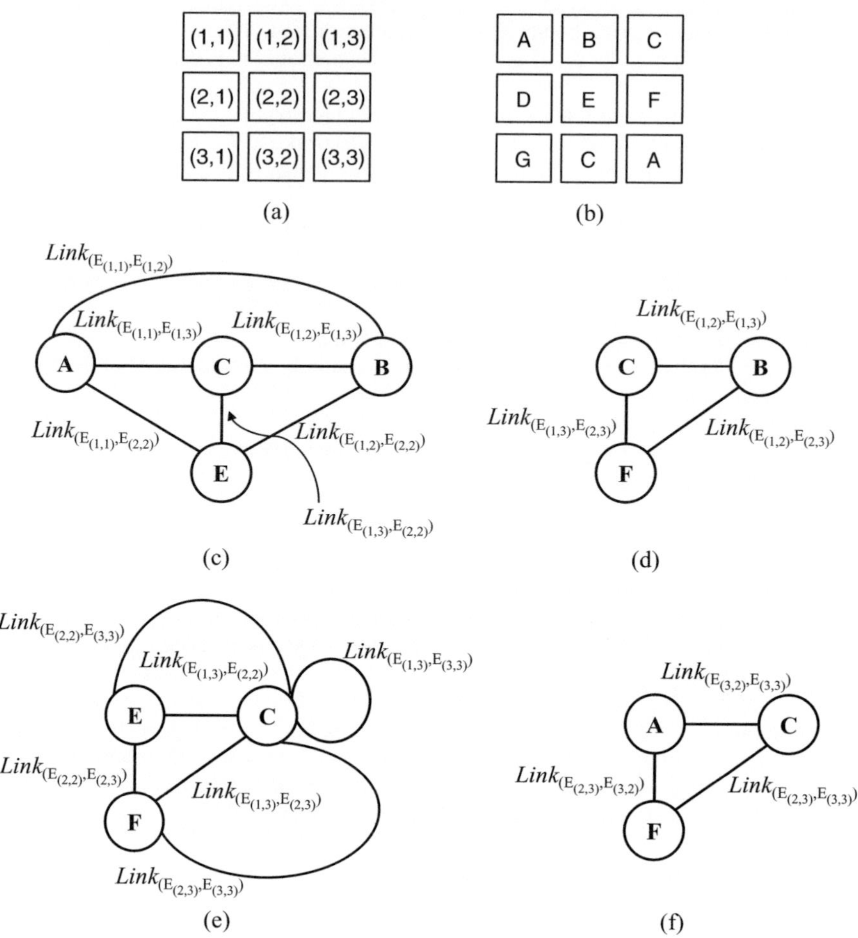

Fig. 6.4 (**a**) Coordinate locations for the electrodes; (**b**) pin-assignment for the electrodes; graphs corresponding to electrode (**c**) $E_{(1,2)}$; (**d**) $E_{(1,3)}$; (**e**) $E_{(2,3)}$; and (**f**) $E_{(3,3)}$

For example, Fig. 6.4a shows the coordinate locations for the electrodes and Fig. 6.4b shows the corresponding pin-assignment configuration. The electrode at position (i, j) is represented by $E_{(i,j)}$, and $E^* = \bigcup_{i,j} E_{(i,j)}$. When the operator $\mathscr{C}$ is applied to $E_{(1,1)}$, we get $\mathscr{C}(E_{(1,1)}) = $ Pin A.

For any electrode $E \in E^*$, a droplet can be virtually placed on this electrode and the CEG of the virtual droplet (written as E_{group}) can be obtained. By taking any two elements from E_{group}, we can get an unordered pair of electrodes. The set that consists of all such electrode pairs is represented by E_{pair}.

Next, the operator Φ is defined as a mapping from the set of electrodes E^* to the set of electrode pairs E_{pair}.

$$\Phi : E^* \rightarrow E_{pair} \tag{6.5}$$

For example, as shown in Fig. 6.4a, when the operator Φ is applied to $E_{(1,3)}$, the following mapping is obtained:

$$\Phi(E_{(1,3)}) = \{(E_{(1,3)}, E_{(1,2)}), (E_{(1,3)}, E_{(2,3)}), (E_{(1,2)}, E_{(2,3)})\}.$$

Using operators $\mathscr{C}$ and Φ defined above, for any electrode $E_{(i,j)}$ on the layout, a graph $G_{(i,j)}$ can be constructed as follows. First, a droplet is placed virtually on the electrode $E_{(i,j)}$ and the corresponding CEG ($E_{group(i,j)}$) is obtained. By applying operator $\mathscr{C}$ to each element of $E_{group(i,j)}$, we get the CPG ($P_{group(i,j)}$) that corresponds to this virtual droplet. Each pin in $P_{group(i,j)}$ is mapped to a node in $G_{(i,j)}$.

By applying operator Φ on E^*, the set of electrode pairs $E_{pair(i,j)}$ can be derived. For each electrode pair (E_x, E_y) in $E_{pair(i,j)}$, we apply operator $\mathscr{C}$ to it and get the corresponding pin pair (P_x, P_y), and add an edge between the two nodes that represent P_x and P_y in the graph. This edge is labeled as $Link_{(E_x,E_y)}$. As (E_x, E_y) is an unordered pair, we consider the edges with labels $Link_{(E_x,E_y)}$ and $Link_{(E_y,E_x)}$ as the same edge, i.e., we do not distinguish between them.

In this way, a one-to-one mapping from the set $E_{pair(i,j)}$ to the edges in $G_{(i,j)}$ is defined. Figures 6.4c-f show graphs $G_{(1,2)}$, $G_{(1,3)}$, $G_{(2,3)}$, and $G_{(3,3)}$, which correspond to electrodes $E_{(1,2)}$, $E_{(1,3)}$, $E_{(2,3)}$, and $E_{(3,3)}$ in Fig. 6.4a, respectively.

If two or more elements in the CEG of $E_{(i,j)}$ share the same control pin, as in the case of electrode $E_{(2,3)}$ shown in Fig. 6.4a, graph $G_{(i,j)}$ will contain cyclic edges, and there will be multiple edges between certain pairs of distinct nodes. Figure 6.4e shows the graph $G_{(2,3)}$ corresponding to electrode $E_{(2,3)}$, which is a multigraph.

Based on above examples, we conclude that $G_{(i,j)}$ is a complete graph if Constraint 1 in Lemma 6.1 is satisfied (i.e., the elements in the CEG of $E_{(i,j)}$ are assigned to different pins). Hence the number of edges in the graph $G_{(i,j)}$ can be derived from the number of nodes in $G_{(i,j)}$ (i.e., the number of elements in the corresponding CEG).

For a given pin-assignment configuration, let the set of electrodes be defined as E_{layout}. By virtually placing a droplet on electrode $E_x \in E_{layout}$, the graph G_{E_x} for any electrode can be derived. The graph for the pin-assignment configuration, which is written as G_{layout}, is defined as the *union* of all G_{E_x} ($E_x \in E_{layout}$), i.e., $G_{layout} = \bigcup\limits_{E_x \in E_{layout}} G_{E_x}$, where the set of nodes in G_{layout} is obtained by applying the union operation to the set of graphs in G_{E_x}, $\forall E_x \in E_{layout}$. In the union graph, edges with the same label are considered as the same edge [12].

Next, we use two examples to illustrate the *union* of two graphs. Figure 6.5a shows the *union* of graphs $G_{(1,2)}$ and $G_{(1,3)}$. Graphs $G_{(1,2)}$ and $G_{(1,3)}$ can be found in Fig. 6.4c and d, respectively. Note that the edge between nodes B and C in $G_{(1,2)}$ and the edge between nodes B and C in $G_{(1,3)}$ are both labeled as $Link_{(E_{(1,2)},E_{(1,3)})}$, hence they are considered to be the same edge in the merged graph. In the merged graph $G_{(1,2)} \cup G_{(1,3)}$, there is only one edge $Link_{(E_{(1,2)},E_{(1,3)})}$ between nodes B and C. We find that the merged graph is a simple graph after checking the number of edges between each pair of nodes.

Fig. 6.5 (**a**) The *union* of graphs $G_{(1,2)}$ and $G_{(1,3)}$; (**b**) the *union* of graphs $G_{(1,2)}$ and $G_{(3,3)}$. The graphs $G_{(1,2)}$, $G_{(1,3)}$, and $G_{(3,3)}$ are shown in Fig. 6.4c, 6.4d, and Fig. 6.4f, respectively

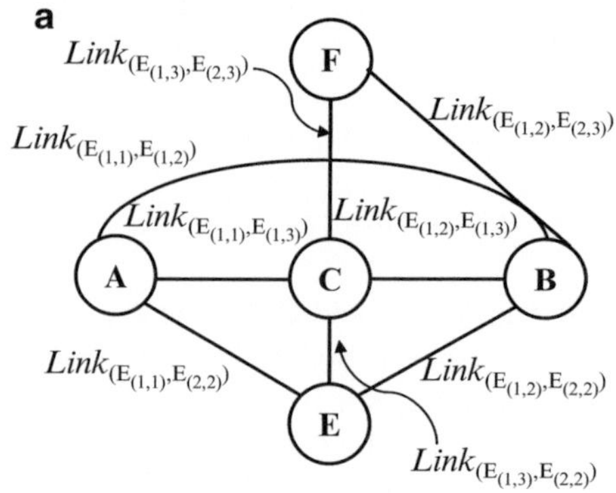

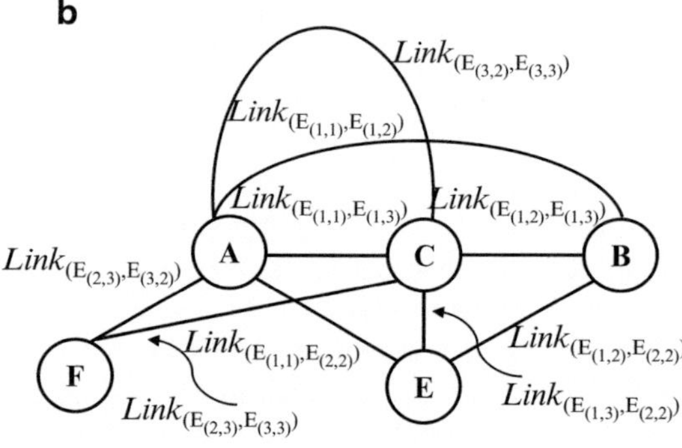

Figure 6.5b is the *union* of graphs $G_{(1,2)}$ and $G_{(3,3)}$. Graphs $G_{(1,2)}$ and $G_{(3,3)}$ can be found in Fig. 6.4c and f, respectively. Note in graph $G_{(1,2)}$ the edge between nodes A and C is labeled as $Link_{(E_{(1,1)},E_{(1,3)})}$, and in graph $G_{(3,3)}$ the edge between nodes A and C is labeled as $Link_{(E_{(3,2)},E_{(3,3)})}$. Therefore the merged graph $G_{(1,2)} \cup G_{(3,3)}$ has two edges between nodes A and C, i.e., $G_{(1,2)} \cup G_{(3,3)}$ is a multigraph.

From the structure of G_{layout} and the acceptability of the pin-assignment (Sect. 6.1), the following lemma can be derived. Note that depending on the mapping of control pins to electrodes, G_{layout} may be a multigraph (either with a self-loop or more than one edge between some pairs of nodes) or a simple graph (no self-loop and no more than one edge between any two distinct nodes) [12].

Lemma 6.2. *A pin-assignment configuration satisfies Constraint 1 and Constraint 2 in Sect. 6.1 if and only if the graph G_{layout} is a simple graph.*

Proof. First we assume that Constraint 1 in Lemma 6.1 is violated. Thus, an electrode group $E_{group(x)}$ exists in which two elements share the same control pin. According to the definition of edges for graph G_{E_x}, there will be cyclic edges and multiple edges between two nodes. Thus G_{E_x} is a multigraph. Since G_{E_x} is a subgraph of G_{layout}, G_{layout} is also a multigraph.   □

We next assume that Constraint 2 in Lemma 6.1 is violated. Thus two non-overlapping CEGs E_{group1} and E_{group2} exist, and their corresponding CPGs share more than one common pin. Assume the graph derived from E_{group1} and E_{group2} are G_1 and G_2, respectively. Suppose $E_{x_1} \in E_{group1}$ and $E_{x_2} \in E_{group2}$ are both connected to pin X, and $E_{y_1} \in E_{group1}$ and $E_{y_2} \in E_{group2}$ are both connected to pin Y. Electrode pair $\{E_{x_1}, E_{y_1}\}$ is mapped to an edge between the nodes that represent pin X and pin Y according to the definition of the one-to-one mapping from the set of electrode pairs to edges in the graph. The label of this edge is $Link_{(E_{x_1}, E_{y_1})}$ (or $Link_{(E_{y_1}, E_{x_1})}$). Electrode pair $\{E_{x_2}, E_{y_2}\}$ is mapped to an edge between the nodes that represent pin X and pin Y. The label of this edge is $Link_{(E_{x_2}, E_{y_2})}$ (or $Link_{(E_{y_2}, E_{x_2})}$). Based on the definition of the *union* of two graphs, when G_1 and G_2 are merged to $G_1 \cup G_2$, there will be two edges with different labels between the nodes that represent pin X and pin Y. Thus $G_1 \cup G_2$ is a multigraph. As $G_1 \cup G_2$ is a subgraph of G_{layout}, we can conclude that G_{layout} also is a multigraph.

According to the definition of G_{layout}, it is easy see that if it is a simple graph, then the pin-assignment configuration satisfies Constraint 1 and Constraint 2.

For any given layout, according to its shape, we can estimate a lower bound on the number of control pins needed to avoid pin-actuation conflicts for any target application.

Here we show two examples for the calculation of the lower bound of the number of control pins.

Theorem 6.1. *Consider an $m \times n$ digital microfluidic array. Then suppose a pin-assignment configuration with M pins exists, such that Constraint 1 and Constraint 2 in Sect. 6.1 are satisfied. A lower bound on M is given by:*

$$\binom{M}{2} \geq 6mn - 5m - 5n + 2.$$

For a $m \times n$ electrode array, and for any electrode $E_{(i,j)}$ in this array, since its corresponding graph $G_{(i,j)}$ is a complete graph, the number of edges can be derived from the number of elements in the CPG. For different positions of the electrodes, the numbers of elements in CPG are different. Thus we count the number of edges G_{layout} by classifying electrode into three categories according to their positions, as shown in Fig. 6.6a.

The first category includes the electrodes located at the corner of the array. The corresponding graph is shown in Fig. 6.6a. For each graph, there are three nodes and the number of edges is $\binom{3}{2} = 3$. Since the number of such electrodes in the $m \times n$ array is 4, we get 4 graphs with 3 edges. The second category includes the electrodes located at four sides but not the corners of the array, and the corresponding graph is shown in Fig. 6.6b. For each graph, there are four nodes and the number of edges is $\binom{4}{2}=6$. As the number for such electrodes is 2(m+n-4), we have 2(m+n-4) graphs with 6 edges. The third category includes the electrodes located within the array and the corresponding graph is shown in Fig. 6.6c. For each graph, there are five

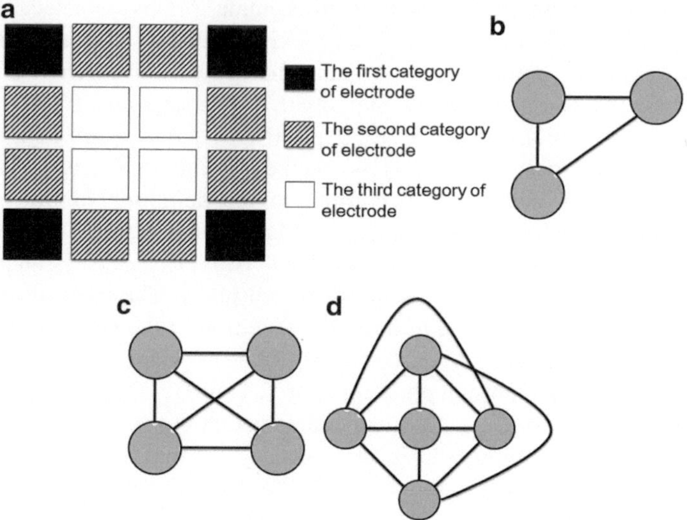

Fig. 6.6 (**a**) Three categories of electrodes and the graphs correspond to electrodes of (**b**) the first category (**c**) the second category (**c**) the third category

nodes and the number of edges is $\binom{5}{2}$=10. Since the number of such electrodes is $(m-2) \times (n-2)$, we will get $(m-2) \times (n-2)$ complete graphs with 10 edges.

On the other hand, the number of edges that are contained in two graphs is $(m-1) \times n + (n-1) \times m + 2 \times (m-1) \times (n-1)$. Using the principle of inclusion/exclusion, we set the total number of edge in the graph G_{layout} to be:

$$N_{edge} = 4 \times 3 + 12(m + n - 4) + 10(mn - 2m - 2n + 4)$$
$$- [(m-1)n + (n-1)m] - 2 \times (m-1) \times (n-1)$$
$$= 6mn - 5m - 5n + 2$$

According to Lemma 6.2, if the pin-assignment configuration satisfies both Constraint 1 and Constraint 2, then graph G_{layout} is a simple graph. Assume that there are M control pins in the pin-assignment configuration. The number of nodes in graph G_{layout} is equal to M. The maximum number of edges in the simple graph G_{layout} that has M nodes is given by:

$$N_{edge\_max} = \binom{M}{2} = \frac{M \times (M-1)}{2}$$

Finally, we derive the desired lower bound in the number of control pins. Since N_{edge_max} is an upper limit on the number of edges in a graph, we have the inequality $N_{edge\_max} \geq N_{edge}$.

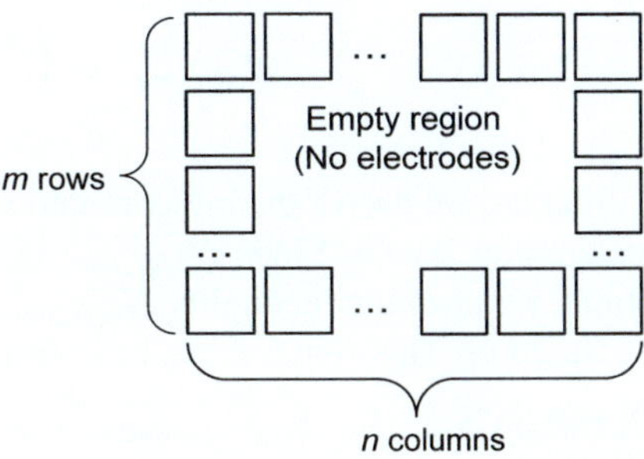

Fig. 6.7 An $m \times n$ outlined rectangle

Based on Theorem 6.1, we have following corollary:

Corollary 6.1. *For large values of m and n, the lower bound on the number of control pins can be approximated as $M_{min} \approx 2\sqrt{3mn}$, and if $m = n = t$, we get $M_{min} \approx 2\sqrt{3}t$. In other words, when the total number of electrodes in the array is N, the number of pins is $\Omega(\sqrt{N})$ (Fig. 6.7).*

Next we calculate the lower bound for the number of control pins required by an $m \times n$ outlined rectangular layout. The layout is shown in Fig. 6.7, and the outlined rectangular array is used extensively in commercial biochips [1, 2]. For the outlined rectangular array, we have the following theorem:

Theorem 6.2. *A lower bound on the number of control pins in the $m \times n$ outlined rectangle is given by following inequality:*

$$\binom{M}{2} \geq 4m + 4n - 8.$$

Proof. For this layout, there are $(2(m + n) - 4)$ CEGs in total and each CEG has three electrodes. When we map the layout to the graph model, we get $(2(m+n)-4)$ complete graphs, and each graph has three nodes and three edges. Some edges are contained in more than one sub-graph. The number of edges that are contained in two graphs is $(2 \times (n - 1)+2 \times (m - 1))$. Therefore, according to the principle of inclusion/exclusion, we get the total number of edges in the graph G_{layout} as:

$$N_{edge} = 3 \times (2(m + n) - 4) - (2 \times (n - 1) + 2 \times (m - 1))$$

$$= 4m + 4n - 8 \qquad \square$$

According to Lemma 6.2, if the pin-assignment configuration satisfies Constraint 1 and Constraint 2, then graph G_{layout} is a simple graph. Assume that there are M control pins in the pin-assignment configuration, the number of nodes in graph G_{layout} is equal to M. The maximum number of edges N_{edge_max} in the simple graph G_{layout}, which has M nodes, is given by:

$$N_{edge\_max} = \binom{M}{2} = \frac{M \times (M-1)}{2}$$

Finally, we derive the lower bound in the number of control pins for the outlined rectangular layout. Since N_{edge_max} is an upper limit on the number of edges in a graph, we have the inequality $N_{edge\_max} \geq N_{edge}$.

Based on Theorem 6.2, we have following corollary:

Corollary 6.2. *For large values of m and n, the lower bound on the number of control pins can be approximated as $M_{min} \approx 2\sqrt{m+n}$, and if $m = n = t$, we get $M_{min} \approx 2\sqrt{2t}$. In other words, when the total number of electrodes in the array is N, the number of pins is $\Omega(\sqrt{N})$.*

For an electrode array that has other shapes, the number of control pins can be lower-bounded in a similar manner.

Let us now return to the pseudocode shown in Fig. 6.3. In the first phase, we determine the lower bound N_L on the number of pins, and then initialize the set of available pins (which is written as *PinAvailable* in Fig. 6.3) as {Pin 1, Pin 2, ..., Pin N_L}. In the second phase, we set an index number for each electrode on the layout by numbering the electrodes one by one, from the electrode on the upper-right corner to the lower-left corner. An electrode E_R is randomly selected as the "seed" of the pin-assignment process, and Pin 1 is assigned to E_R. The set of electrodes whose control pins have been determined is written as "ElectrodesAlreadyAssigned" (as shown in Fig. 6.3). The graph corresponding to the pin-assignment configuration of ElectrodesAlreadyAssigned is written as G_{EAA}.

Then electrode E_R is the first element that is added into the set ElectrodesAlreadyAssigned. The pin-assignment configuration for the layout is constructed electrode-by-electrode around the seed E_R. In each step, we first locate all the neighboring electrodes of ElectrodesAlreadyAssigned. Then these electrodes are sorted out to identify the one that has the minimum index number. The electrode with the minimum index number is written as $E_{neighbor}$ as shown in Fig. 6.3. Then, starting from Pin 1 and proceeding through Pin N_L, we search for the control pin that can be assigned to $E_{neighbor}$ to obtain a feasible pin-assignment solution. Here are the detailed steps of the searching process.

Each time we virtually assign a control pin to $E_{neighbor}$, and update G_{EAA} by doing the union of the original G_{EAA} with the graph that corresponds to the new electrode. Then we determine whether the updated G_{EAA} is a simple graph. Based on Lemma 6.2, if G_{EAA} is a simple graph, then we have expanded the feasible local solution successfully and $E_{neighbor}$ will be added into the set ElectrodesAlreadyAssigned, as shown in Fig. 6.3. Otherwise, the pin that is assigned to $E_{neighbor}$ will be changed.

If none of the pins in set *PinAvailable* can be assigned to $E_{neighbor}$, a new pin is added into *PinAvailable*, and this newly added pin is assigned to $E_{neighbor}$. By repeating the above procedure, the set ElectrodesAlreadyAssigned can be expanded step-by-step until it covers the whole layout. When we assign pins to the newly-added electrode, the *complement* of the graph G_{EAA}, referred to as H_{EAA}, is used to reduce the search scope. All the pins corresponds to isolated nodes in H_{EAA} are "unavailable pins" and they cannot be assigned to E_{neighbor}.

Assume that the number of electrodes on the array is $\mathcal{N}$ and the number of pins assigned to the electrodes is $\mathcal{P}$. The computing complexity for checking whether G_{EAA} is a simple graph is $\mathrm{O}(\mathcal{P}^2)$. Each time when we assign a pin to E_{neighbor}, we have to verify whether G_{EAA} is a simple graph. Thus the computational complexity of assigning a pin to E_{neighbor} is $\mathrm{O}(\mathcal{P}^3)$ for the worst case. Since all electrodes are added step-by-step to the set ElectrodesAlreadyAssigned, the overall computational complexity of the heuristic procedure is $\mathrm{O}(\mathcal{P}^3 \mathcal{N})$.

The heuristic algorithm can therefore quickly generate a feasible pin-assignment configuration, and for each given array layout, the number of control pins derived by the heuristic algorithm can be used as the upper bound on the number of control pins. For $n \times n$ rectangular layouts and $n \times n$ outlined rectangular layouts, Fig. 6.8a and b show the lower bounds for the number of pins derived from the theoretical analysis and the upper bounds for the number of pins derived from the heuristic algorithm.

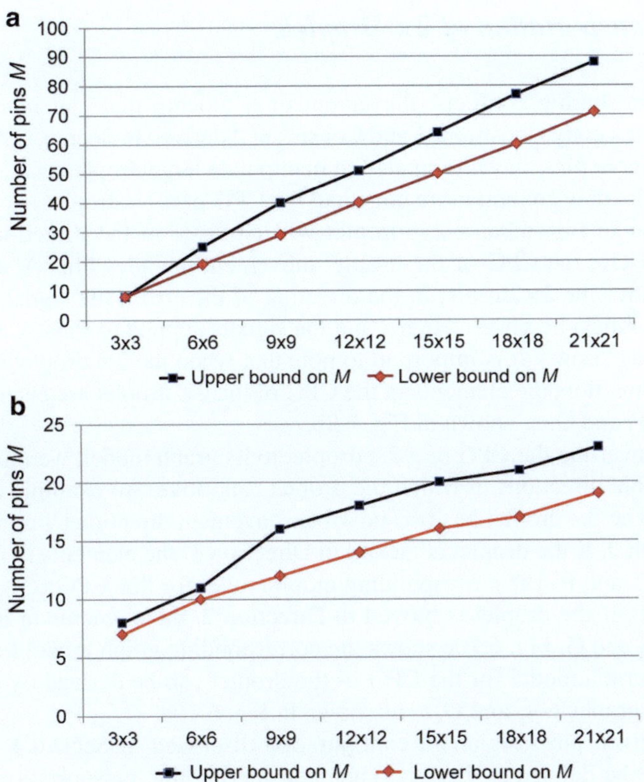

Fig. 6.8 The relationships between lower/upper bounds on pin-counts and the sizes of the electrode arrays for (**a**) rectangular layouts; (**b**) outlined rectangular layouts

6.5   Manipulation of Large Droplets

Bioassay benchmarks used to evaluate biochip design automation methods in the literature make the assumption that all input and output droplets of mixing/dilution operations are 1× droplets [13, 14]. However, with recent advances in fabrication technology, digital microfluidic biochips can now transport 2× and even larger droplets; the transportation of larger droplets requires higher actuation voltages [15]. Some newly developed bioassay protocols for lab-on-chip include the manipulation of 2× droplets. For example, to capture antibodies, a 1× blood droplet needs to be mixed with a 2× droplet that contains magnetic beads in the "on-magnet incubation" protocol [2]. In order to execute this bioassay, the biochip needs to be able to transport 2× and even larger droplets.

To move droplets with large volumes, more electrodes need to be activated at each clock cycle [16]. However, the manipulation of droplets with different volumes introduces new constraints for the design of pin-limited biochips. We discuss these details below.

6.5.1   Transportation of 2× Droplets

To avoid pin-sharing conflicts, the layout of a biochip that can manipulate 2× droplets must satisfy Constraint 1 and Constraint 2 derived in Sect. 6.1. A difference here is that more pins may be actuated to manipulate large droplets, i.e., the CPG of a large droplet may contain more pins than the CPG of a 1× droplet.

As shown in Fig. 6.9a, a 2× droplet will elongate in the direction of movement [16]. Here, the CPG of the droplet movement includes Pins A, B, C, D, E, and F. To move the 2× droplet in the direction of the arrow, the signals applied to pins B and C must be set as "High", and the signals applied to Pins A, D, E, and F must be set as "Low". It is important to note that when the 2× droplet is moved to a different direction, the elements in the CPG of the 2× droplet are changed to Pins A, B, D, E, F, and G, as shown in Fig. 6.9b.

When converting the CPG of a 2× droplet to its graph model, we must consider all the possible directions in which the droplet can move. An example is shown in Fig. 6.10a. The 2× droplet has two possible movement directions, i.e., Direction 1 and Direction 2. If the droplet is moved in Direction 1, the elements in its CPG are Pins A, B, C, and F. The corresponding graph model for this CPG (G_{D_1}) is shown in Fig. 6.10b; if the droplet is moved in Direction 2, the elements in its CPG are Pins A, B, F, and G. Fig. 6.10c shows the corresponding graph model for this CPG (G_{D_2}). The graph model for the CPG of the droplet can be derived by calculating the *union* of graphs G_{D_1} and G_{D_2}, as shown in Fig. 6.10d.

For the given pin-assignment configuration discussed in Sect. 6.4, let the set of electrodes be defined as E_{layout}. When considering the transportation of the 2× droplet, we can construct a graph model for the layout E_{layout} by virtually placing a

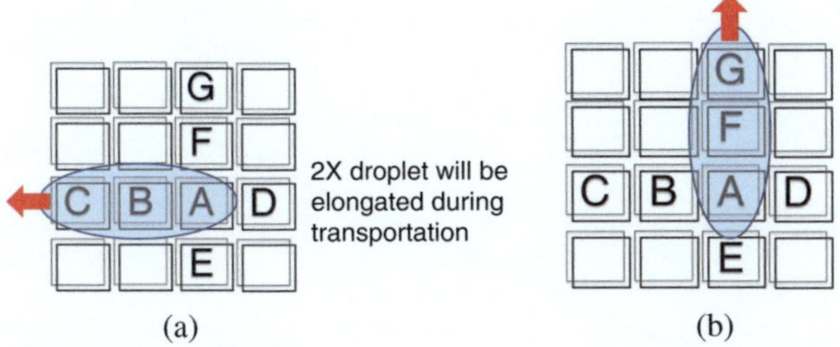

Fig. 6.9 (**a**) To move the 2× droplet to the arrow's direction, Pins B and C must be "High", and Pins A, D, E, and F must be "Low"; (**b**) to move the droplet in another direction, pins F and G must be "High", and Pins A, B, D, and E must be "Low"

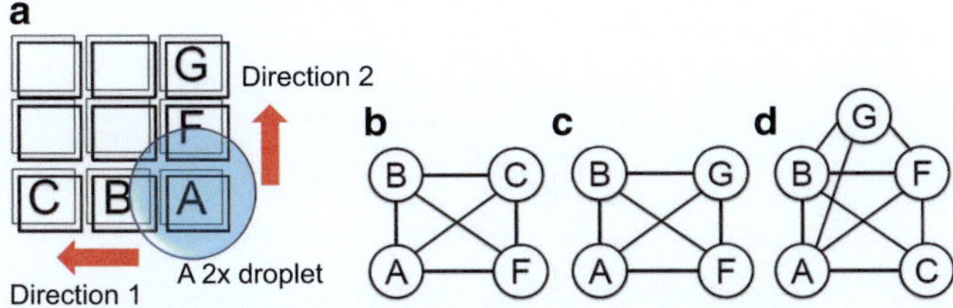

Fig. 6.10 (**a**) The 2× droplet has two possible movement directions; (**b**) G_{D_1}: graph model of CEG corresponding to the movement in Direction 1; (**c**) G_{D_2}: graph model of CEG corresponding to the movement in Direction 2; (**d**) graph model derived by the union of G_{D_1} and G_{D_2}

2× droplet on electrode $E_x \in E_{layout}$. As discussed above, when the 2× droplet is moved in different directions, the elements in the CPG are different. Note that when we design the pin-assignment configuration, we do not assume any knowledge of the position of the 2× droplets or the directions in which they must be moved in different clock cycles. Therefore, 2× droplets can be moved in any non-diagonal direction, and these 2× droplets can be present at any position on the layout. To derive the graph G_{E_x} for any electrode, we must consider all the possible directions in which the droplet can move, and obtain their corresponding graphs as explained below.

The graph for the pin-assignment configuration, which is written as G_{layout}, is defined as the *union* of all the G_{E_x} graphs, i.e., $G_{layout} = \bigcup_{E_x \in E_{layout}} G_{E_x}$, where G_{E_x} is defined as: $G_{E_x} = \bigcup \{G_{E_x D_1}, G_{E_x D_2}, \ldots G_{E_x D_D}\}$, and $D_1, D_2, \ldots, D_D$ are the possible directions for the movement of the 2× droplet that stays on electrode E_x.

We can easily prove that, for the biochips involving the transportation of 2× droplets, Lemma 6.2 can still be used as an acceptance test for feasible pin-assignment configurations. The heuristic algorithm shown in Fig. 6.3 can be used for the biochip with manipulations of 2× droplets. Similarly, we can determine the upper and lower bounds for the number of control pins.

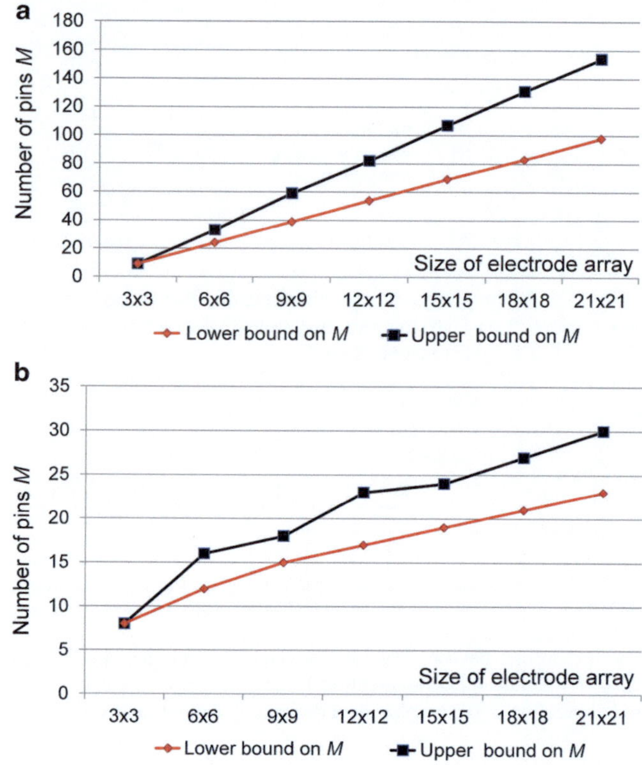

Fig. 6.11 (**a**) Relationship between lower/upper bounds on pin-counts and the sizes of the electrode array (rectangular layouts) for biochips with 2× droplets; (**b**) relationship between lower/upper bounds on pin-counts and the sizes of the electrode array (outlined rectangular layouts) for biochips with 2× droplets

For an $n \times n$ array, the lower bound and upper bound for the number of control pins are shown in Fig. 6.11a. For an $n \times n$ outlined rectangular array, the lower bound and the upper bound for the number of control pins are shown in Fig. 6.11b. By comparing the lower/upper bounds on the number of control pins presented in Fig. 6.8 and 6.11, we can find that the number of pins assigned to the biochip has to increase in order to manipulate 2× droplets.

6.5.2 Influence of Diagonal Electrodes

In this subsection, we discuss the influence of diagonally adjacent electrodes on the movement of a droplet. Assume that the droplet shown in Fig. 6.12a will be moved in the direction of the arrow, and the electrode that the droplet will be moved to is actuated. The contact line between the droplet and the actuated electrode is shown

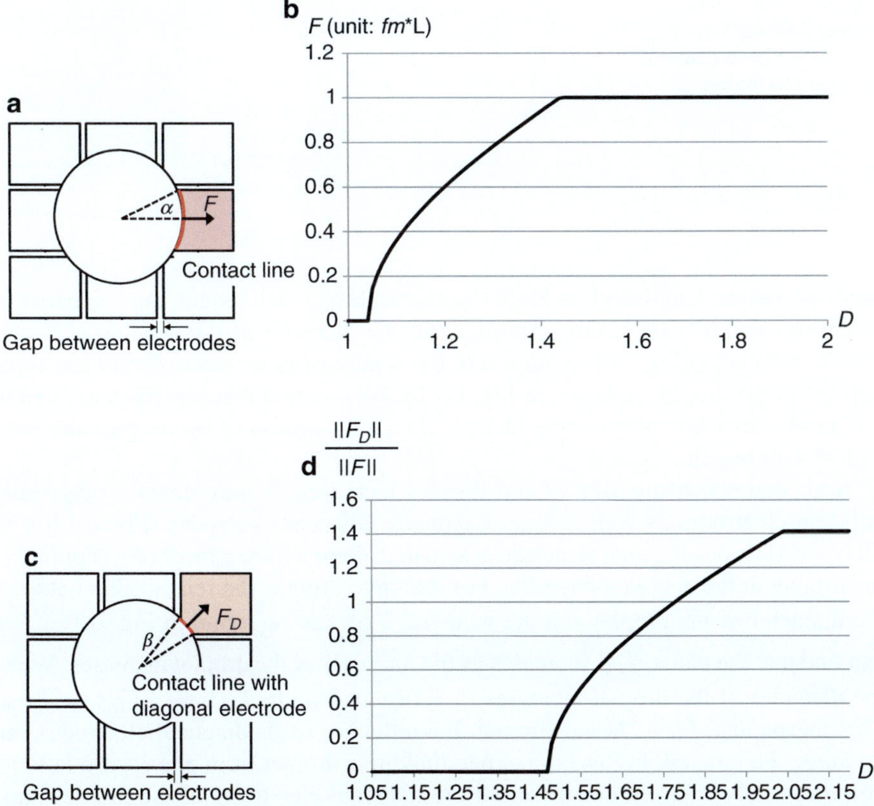

Fig. 6.12 (**a**) The force applied to the droplet when a non-diagonally adjacent electrode is actuated; (**b**) relationship between the diameter of the droplet and the force F applied to the droplet; (**c**) the force F_D applied to the droplet when a diagonally adjacent electrode is actuated; (**d**) relationship between the diameter of the droplet and $\frac{\|F_D\|}{\|F\|}$

in Fig. 6.12a. According to the principle of EWOD, the total electrowetting force F applied to the droplet can be written as [16]:

$$F = f_m \cdot L \cdot sin\alpha, \tag{6.6}$$

where L is the diameter of the droplet, and the angle α is shown in Fig. 6.12a [16]. The parameter f_m is the driving force per unit length of the contact line [16], which is determined by the surface characteristics of the device and the actuation voltage applied to the electrode. For a given liquid in the droplet and a given biochip, f_m can be considered as a constant [16].

For a typical EWOD device, the size of the square electrodes is 1×1 mm. As shown in Fig. 6.12a and c, there is a gap between the two adjacent electrodes. The gap between two adjacent electrodes in the typical EWOD device is 20 μm [17].

Fig. 6.13 The "large"
droplet contacts both
diagonal and non-diagonal
adjacent electrodes

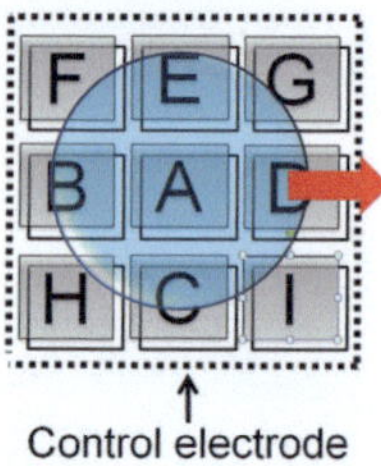

So, if we set the length and width of the electrode as 1 unit length, the gap between electrodes is 0.02 unit length. Assume that the diameter of the droplet is D unit length, then according to Equation 6.6, the relationship between D and the force applied to the droplet is shown in Fig. 6.12b. We can find that, the force applied to the droplet does not increase any further when the diameter of the droplet increases to 1.46 unit length.

Note that when the size of the droplet increases, it may contact diagonally adjacent electrodes as well as non-diagonally adjacent electrodes (Fig. 6.12c). In this case, if a non-diagonal electrode is actuated, there will be a force (F_D) applied to the droplet in the diagonal direction. For the same droplet, the relationship between the diameter of the droplet and the value of $\frac{\|F_D\|}{\|F\|}$ can be found in Fig. 6.12d. We can find that the value $\frac{\|F_D\|}{\|F\|}$ increases as the diameter of the droplet increases. When the diameter of the droplet increases to 1.78 unit length, we have $\|F_D\| = \|F\|$. This means that, for a "large" droplet, the influence of its diagonal electrodes can no longer be ignored. In this case, when moving a droplet from one electrode to its non-diagonally adjacent electrode, all the diagonal electrodes of the droplet must be applied "Low" voltage. Otherwise these diagonal electrodes of the droplet will influence the movement of the droplet.

An example is shown in Fig. 6.13. In this case, the movement of the droplet is influenced by the nine electrodes in contact with it. In order to move the droplet in the direction of the arrow, pin D must be "High", while the low voltage must be applied to all other pins. From this example, we can find that biochips that are required to manipulate a large droplet must satisfy the Constraint 1 and Constraint 2 in Lemma 6.1 in Sect. 6.1 with the difference that the CEG includes all the nine electrodes in contact with the droplet.

6.6   Scheduling of Fluid-Handling Operations

For the proposed general-purpose biochip, fluid-handling operations must be scheduled after the pins are assigned to electrodes. This is in contrast to bioassay-specific design where the scheduling precedes pin-assignment. In our work the initial schedule is derived without considering any pin constraints, and then the schedule is adjusted according to the pin-assignment configuration.

For example, when multiple fluid-handling operations are scheduled to be executed concurrently, their control signals may conflict with each other. In order to solve this problem, we must decide the order in which the fluid-handling operations will be implemented and determine the appropriate voltages to be applied to the electrodes.

Suppose we have a given initial schedule. Then at each time slot t, the positions and scheduled fluid-handling operations of all the droplets are already known. Therefore, the corresponding actuation signals that need to be applied on the pins can be easily derived. Based on the actuation signals, the set of pins that must be applied "High" signal ($S_H(t)$), and the set of pins that must be applied "Low" signal ($S_L(t)$) can be determined. Conflicts in the actuation of the pins can occur if and only if $C(t) = S_H(t) \cap S_L(t) \neq \varnothing$, i.e., the operation schedule requires some control pins to be set to "High" and "Low" at the same time. In this situation, the operation schedule needs to be adjusted.

An example is shown in Fig. 6.14a. Assume at time $t = t_0$, droplets D_1, D_2, and D_3 are scheduled to be moved in the directions indicated by the arrows, and D_4 is scheduled to be split. To implement the fluid-handing operation for droplets D_3, we have the following set of pins that must be applied "High" voltage: $S_H^{D_3}(t_0) = \{11\}$; to implement the fluid-handing operation for droplets D_4, we have the following set of pins that need to be applied "Low" voltage: $S_L^{D_4}(t_0) = \{2, 11, 14\}$. Hence we have $S_H^{D_3}(t_0) \cap S_L^{D_4}(t_0) \neq \varnothing$, i.e., the fluid-handling operations scheduled for droplets D_3 and D_4 cannot be implemented concurrently, otherwise the actuation signals for D_3 may lead to unintended movement of D_4. Note that the moving and splitting of droplets are the basic operations implemented on the biochip, and other operations such as droplet mixing and dispensing can be considered as a sequence of these basic operations.

To determine the actuation signals that must be applied to the electrodes and the order in which to implement the fluid-handling operations, we map the conflict relationships of multiple droplets to a directed graph, and each droplet is mapped to a node. For two droplets D_A and D_B, if the fluid-handling operation of D_A may lead to unintended splitting or movement of D_B, then in the conflict graph, there will be a directed edge starting from the node D_A to node D_B. The operations for droplets that correspond to the nodes with zero out-degrees in the conflict graph must be first

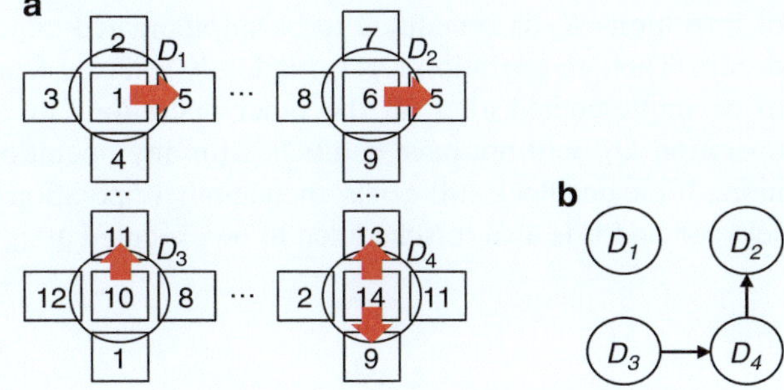

Fig. 6.14 (a) Droplets D_1, D_2, and D_3 are scheduled to be moved in the directions indicated by the arrows, and D_4 is scheduled to be split; (b) conflict graph derived from the pin-assignment configuration and scheduled fluid-handling operations of droplets D_{1-4}

executed, as these operations will not result in unintended splitting or movement of other droplets.

Figure 6.14b shows the conflict graph derived from the scheduled fluid-handling operations of droplets shown in Fig. 6.14a. According to the conflict graph, we can see that droplets D_1 and D_2 can be moved at time $t = t_0$, because their fluid-handling operations will not lead to unintended splitting or movement of other droplets. Droplets D_3 and D_4 must remain at their current positions.

Deadlock in fluid-handling operations occurs if the manipulation of any droplet may lead to unintended splitting or movement of one or more other droplets. In this case, none of the fluid-handling operations can be executed. We can find the following necessary and sufficient condition for the occurrence of deadlock from the conflict graph.

Lemma 6.3. *Deadlock occurs at time t if and only if all the nodes in the corresponding conflict graph have non-zero out-degrees.*

Proof. First, assume that deadlock occurs and none of the scheduled fluid-handling operations can be executed. Hence, executing the operation of an arbitrarily-chosen droplet D_K may lead to the unintended splitting or movement for a group of other droplets $\{D_{K_1}, D_{K_2}, \ldots, D_{K_n}\}$. Based on the definition of edges in the conflict graph, directed edges will start from the node that represents droplet D_K and go to the nodes that represent droplets $\{D_{K_1}, D_{K_2}, \ldots, D_{K_n}\}$. The out-degree of the node that represents droplet D_K is non-zero. Since D_K is an arbitrarily-chosen droplet, we can conclude that when deadlock occurs, all the nodes in the conflict graphs have non-zero out-degrees. $\qquad\square$

Next, we prove that Lemma 6.3 provides a sufficient condition for the occurrence of deadlock. Based on the definition of a directed edge in the conflict graph, if the out-degree for a node is non-zero, moving the droplet that is represented by the node may lead to unintended splitting or movement of other droplets. If all the nodes in the conflict graph have non-zero out-degrees, we can conclude that the manipulation of any droplet may lead to unintended splitting or movement of other droplets. Hence the deadlock occurs.

Based on the above argument, the necessity and sufficiency of the condition in Lemma 6.3 are proven.

When the deadlock occurs, one or more fluid-handling operations need to be re-scheduled. Here are the details of the re-scheduling procedure. Assume that a set of operations S_d is scheduled to be implemented concurrently, and the deadlock occurs. Then an operation O_R is randomly selected from S_d and it is re-scheduled to be implemented after all the other operations in S_d are finished. Therefore, operation O_R will not have conflicts with any operations in the set $S_d \setminus \{O_R\}$ any more. If the deadlock still exists, then another operation from $S_d \setminus \{O_R\}$ is randomly selected and it is also re-scheduled to be executed after all the other operations are finished. This re-scheduling process is repeatedly implemented until the deadlock no longer occurs. The existence of a schedule without the deadlock is guaranteed by the following lemma:

Lemma 6.4. *Deadlock never occurs when there are only two droplets concurrently manipulated by the proposed general-purpose pin-limited biochip.*

Proof. This lemma can be proven by *reductio ad absurdum*. First we assume that at time $t = t_d$, deadlock occurs for two droplets D_x and D_y. According to the definition of deadlock, we know the operation on D_x may lead to the unintended splitting or movement for D_y, i.e., $S_H^{D_x}(t_d) \cap S_L^{D_y}(t_d) \neq \emptyset$. Then we know the operation on D_y may lead to the unintended splitting or movement for D_x, i.e., $S_H^{D_y}(t_d) \cap S_L^{D_x}(t_d) \neq \emptyset$. $\square$

For each control pin of droplet D_x, it is either in the set of $S_H^{D_x}(t_d)$ or in the set of $S_L^{D_x}(t_d)$. So we have $S_H^{D_x}(t_d) \cap S_L^{D_x}(t_d) = \emptyset$, and the control pin group of D_x can be written as $\mathrm{CPG}_{D_x}(t_d) = S_H^{D_x}(t_d) \cup S_L^{D_x}(t_d)$.

Similarly, for droplet D_y, we have $S_H^{D_y}(t_d) \cap S_L^{D_y}(t_d) = \emptyset$, and the control pin group of D_y can be written as $\mathrm{CPG}_{D_y}(t_d) = S_H^{D_y}(t_d) \cup S_L^{D_y}(t_d)$.

Therefore, we have: $\mathrm{CPG}_{D_x}(t_d) \cap \mathrm{CPG}_{D_y}(t_d) = (S_H^{D_x}(t_d) \cap S_L^{D_y}(t_d)) \cup (S_H^{D_y}(t_d) \cap S_L^{D_x}(t_d))$. It is already known that $S_H^{D_x}(t_d) \cap S_L^{D_y}(t_d) \neq \emptyset$ and $S_H^{D_y}(t_d) \cap S_L^{D_x}(t_d) \neq \emptyset$, we conclude that the CPGs of droplet D_x and D_y have at least two common elements. This is in conflict with Constraint 2 in Sect. 6.1. This completes the proof.

Lemma 6.4 shows that in the worst case, we can reduce the number of concurrent fluid-handling operations to two, and deadlock will not occur. Therefore, by re-scheduling the fluid-handling operations, we can always implement the bioassay without deadlock.

The pseudocode for the scheduling algorithm based on pin-actuation sequences can be found in Fig. 6.15. Note the scheduling algorithm can be applied to any pin-limited biochip, including the bioassay-specific biochips proposed in [9, 18].

6.7 Simulation Results

In this section, we present results for two commercial biochips and several laboratory prototypes.

1: Derive initial operation schedule without considering of pin-constraints;
2: Derive the sets of pins $S_H(t)$ and $S_L(t)$ at time t;
3: **if** $S_H(t) \cap S_L(t)$ **then**
4:    Derive the conflict graph of droplets and determine the priorities of fluid-handling operations for droplets;
5:    **if** operations for set of droplets $F(t)$ can be executed without leading to unintended splitting or movement of other droplets **then**
6:       Execute fluid-handling operations for droplets in set $F(t)$;
7:    **else**
8:       Adjust the operation schedule; // Deadlock occurs and operations need to be rescheduled.
9:    **end if**
10: **else**
11:    Execute fluid-handling operations concurrently based on initial scheduling.
12: **end if**

Fig. 6.15 Pseudocode for determining the priorities of the movements of the droplets

6.7.1 Commercial Biochips

6.7.1.1 Commercial Biochip for *n*-plex Immunoassay

In an n-plex immunoassay, a sample is concurrently analyzed for n different analytes. Sample droplets are mixed with n different reagents, and the mixed product droplets are moved to the detection area which includes an optical detector [1, 2]. The commercial biochip for the n-plex immunoassay consists of more than 1000 electrodes [1, 2]. The layout of this biochip is shown in Fig. 6.16a.

As seen in Fig. 6.16a [2, 13], the commercial chip for n-plex immunoassay is based on a regular design that consist of two types of unit cells. We design the unit cells in the layout and assign control pins for each unit cell; all other cells can be assigned pins based on the same configuration. By applying the heuristic algorithm proposed in Sect. 6.4, we obtain the pin-assignment for the Type I and Type II unit cells, as shown in Fig. 6.16b and Fig. 6.16c. Note here these two unit cells are controlled by independent sets of pins, thus pins 1 to 7 are assigned to Type I unit cell, and pins 8 to 18 are assigned to Type II unit cell.

Table 6.1 compares the existing pin-assignment configuration for the fabricated commercial biochip [13], the result derived by bioassay-specific algorithm in [13], and the result derived by the proposed heuristic method. The proposed heuristic method leads to comparable or smaller number of control pins than both [13] and the fabricated commercial biochip. Since the pin-assignment derived by the proposed method is bioassay-independent, it also provides the added benefit of being flexible for multiple target applications.

The CPU time for designing the layout for the commercial biochip can be found in Table 6.1. The existing pin-assignment configuration is derived manually, so it does not have CPU time [19]. The CPU time needed to generate the pin-assignment configuration in [19] is 130 min. while the CPU time is reduced to 7.4 s for the proposed heuristic algorithm.

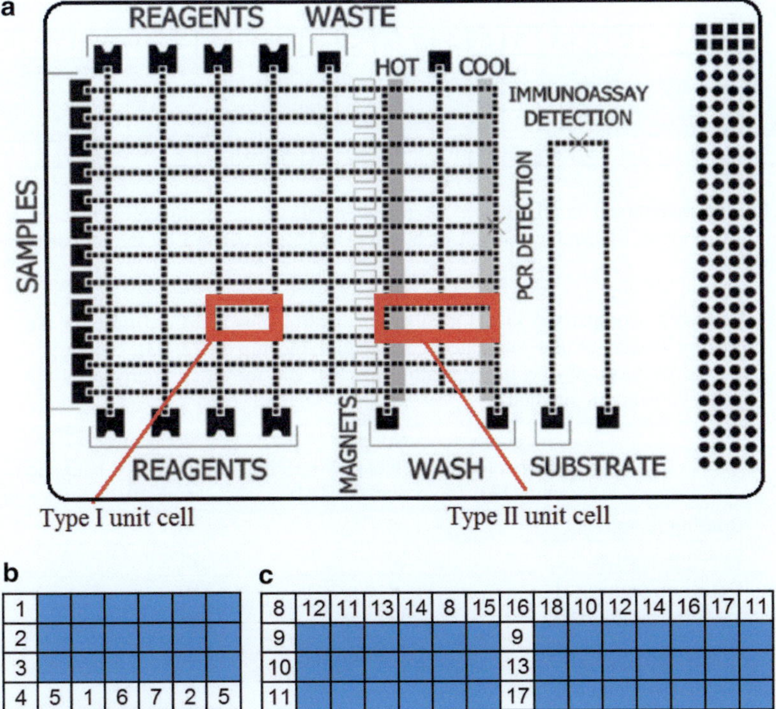

Fig. 6.16 (**a**) Layout of the commercial chip [2, 13] and pin-assignment for the unit cells of (**b**) Type I, and (**c**) Type II shown in Fig. 6.16

Table 6.1 Comparison of the numbers of control pins and CPU time for the existing design of the fabricated commercial biochip in [13], the results derived by bioassay-specific algorithm in [13], and the results derived by the proposed heuristic method

	No. pins (fabricated chip [13])	No. pins (bioassay-specific method [13])	No. pins (proposed heuristic method)
Type I unit cell	7	6	7
Type II unit cell	19	13	11
Total No. of pins	26	19	18
CPU time	NA	130 min [19]	7.4 s

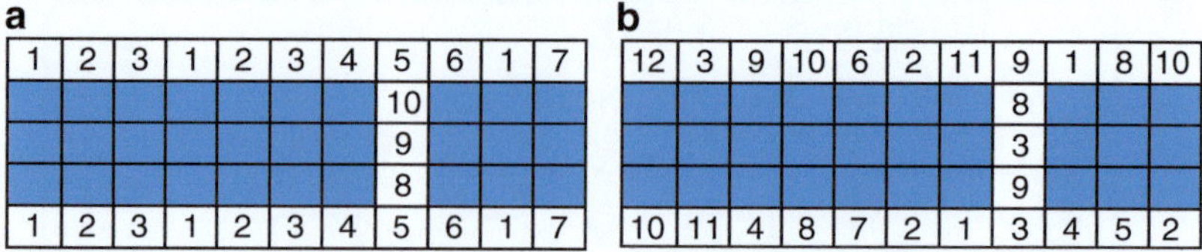

Fig. 6.17 Pin-assignment configuration for the unit cell of the commercial biochip for PCR bioassay: (**a**) derived by synthesis result of bioassay [13]; (**b**) derived by the proposed heuristic algorithm

Table 6.2 Comparison of the numbers of control pins and CPU time for the existing design of the fabricated commercial biochip in [2, 13], the results derived by bioassay-specific algorithm in [13], and the results derived by the proposed heuristic method

	No. pins (fabricated chip [13])	No. pins (bioassay-specific method [13])	No. pins (proposed heuristic method)
Routing region	7	6	6
Reaction region	14	10	14
Detection region	11	8	10
Total No. of pins	32	24	30
CPU time	NA	320 min [19]	3.5 s

6.7.1.2 Commercial Biochip for PCR Bioassay

We next apply the proposed algorithm to the commercial biochip for PCR bioassay [13, 19]. The chip consists of more than 1000 electrodes [13, 19]. The layout for the complete chip and the existing pin assignment in the fabricated biochip (14 pins) can be found in [13, 19]. Pin-assignment configurations for the unit cell of the biochip derived by bioassay-specific algorithm is shown in Fig. 6.17a [13], and the layout derived by the proposed algorithm is shown in Fig. 6.17b.

Table 6.2 compares the number of control pins in the existing design for the fabricated commercial biochip [13], the number of pins derived by bioassay-specific algorithm in [13], and the number of pins derived by the proposed heuristic method. Similar with the results of n-plex immunoassay chip, the CPU time of the proposed heuristic algorithm is 3.5 s, while the CPU time needed for generating the pin assignment in [19] is 320 min.

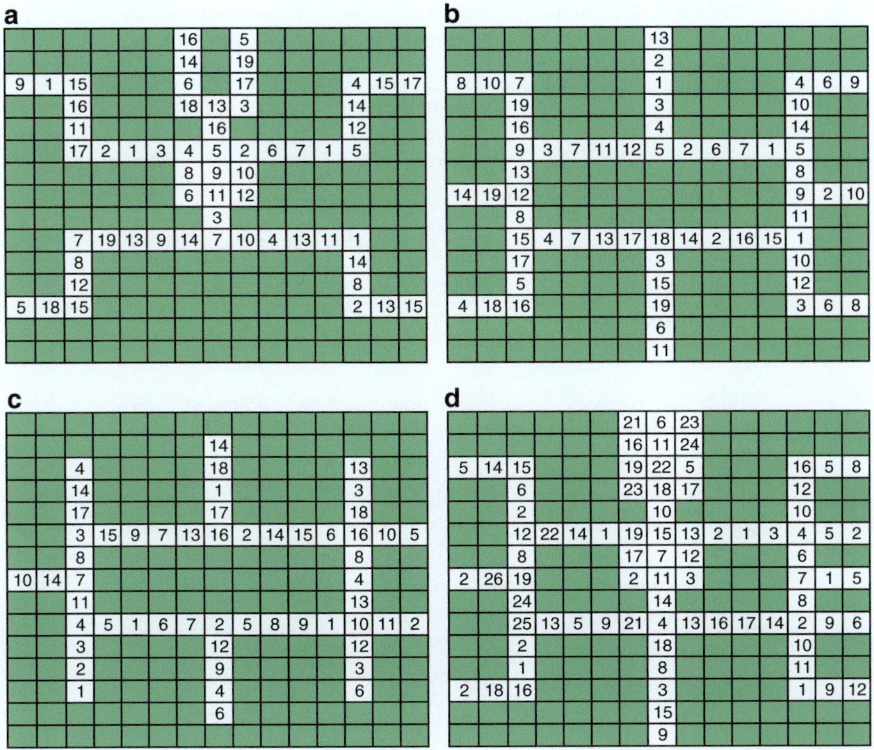

Fig. 6.18 Pin-assignment configurations for (**a**) multiplexed assay chip; (**b**) PCR chip; (**c**) protein dilution chip; (**d**) multi-functional chip

6.7.2 *Experimental Biochips*

Next we apply the pin-assignment algorithm to the layouts of four laboratory prototypes described in [9], namely a multiplexed assay biochip, a PCR biochip, a protein dilution biochip, and a multifunction biochip. These layouts have subsequently been incorporated in commercial chips [20]. Figure 6.18 shows their pin-assignment configurations.

First we compare the number of pins and bioassay completion time between the proposed general-purpose biochips and other biochips, which include the cross-referencing biochip in [6], the bioassay-specific biochips in [9] and [18], and the direct-addressing biochips. We implemented the ILP-based design method of [18] to obtain a complete set of simulation results. Tables 6.3 and 6.4 present the simulation results.

Note that both the cross-referencing biochip in [6] and the proposed general-purpose biochip are bioassay-independent. When we map bioassays to bioassay-independent chips, droplet operations are scheduled after the

Table 6.3 The numbers of pins for the proposed pin-limited biochip, bioassay-specific biochips [9], cross-referencing biochips [6], and direct-addressing biochip

	Multiplexed bioassay No. Pins	PCR No. Pins	Protein bioassay No. Pins
Proposed heuristic method	19	19	18
Cross router [6, 9]	30	30	30
Direct-addressing	59	62	52
Bioassay-specific [9]	25	14	27
ILP-based [18]	11	6	12

Table 6.4 Bioassay execution time on the proposed pin-limited biochip, bioassay-specific biochips [9], cross-referencing biochips [6], and direct-addressing biochip

	Multiplexed bioassay Time (s)	PCR Time (s)	Protein bioassay Time (s)
Proposed heuristic method	63	22	180
Cross router [6, 9]	132	26	195
Direct-addressing	70	19	179
Bioassay-specific [9]	70	20	216
ILP-based [18]	73	19	161

pin-assignment configuration is determined. Therefore, the operation schedule and routing pathways are more flexible, and they can be adjusted according to the requirements of bioassays. Tables 6.3 and 6.4 show that the proposed method can achieve fewer number of pins and shorter execution time, compared with that cross-referencing biochips in [6]. The proposed general-purpose biochip can greatly reduce the number of control pins with only a slight increase in the bioassay completion time, compared with the direct-addressing biochips.

In order to compare the proposed method with the bioassay-specific biochips in [9], we first derive the synthesis results of bioassays without considering pin-assignment configurations. The initial synthesis results can be derived from previously published synthesis tools in [21–23]. Then the operations schedules are adjusted based on the signals applied to pins, as introduced in Sect. 6.6. Tables 6.3 and 6.4 show that, the proposed method can achieve fewer number of control pins and shorter execution time for multiplexed bioassay and protein dilution bioassay, compared with the bioassay-specific biochips in [9].

In the following discussion, the multiplexed assay biochip, PCR bioassay biochip, protein dilution biochip, and the multi-function biochip presented in Fig. 6.18 are referred to as Chips 1–4; these four bioassay-specific biochips designed

Table 6.5 Execution time of bioassays on Chips $1 - 4$

Biochip Bioassay	Chip 1	Chip 2 (s)	Chip 3	Chip 4 (s)
Multiplexed bioassay	63 s	65	NA	62
PCR bioassay	NA	22	NA	23
Protein bioassay	195 s	221	180 s	217

Table 6.6 Execution time of bioassays on Chips $A - D$

Biochip Bioassay	Chip A	Chip B (s)	Chip C	Chip D (s)
Multiplexed bioassay	70 s	77	NA	73
PCR bioassay	NA	20	NA	20
Protein bioassay	264 s	441	220 s	223

in [9] are referred to as Chips $A - D$; and these four biochips derived by ILP-based pin-count aware design in [18] are referred to as Chips $E - G$.

By simulating multiplexed bioassay, PCR bioassay, and protein dilution bioassay on Chips $1-4$, we can evaluate the flexility of the proposed general-purpose biochip. The completion times of these bioassays can be found in Table 6.5. Note even though the Chips 1-4 are general-purpose biochips, their applications are still limited by their available input/output ports. For example, Chip 1 has only 6 input/output ports, while the PCR bioassay requires at least 8 input ports (because it has 8 different input reagents). Hence the PCR bioassay cannot be executed on Chip 1, and the corresponding execution time is not available (written as "NA" in the table). From the simulation results presented in Table 6.5, we can find that the proposed general-purpose biochip can be adapted to different bioassays with only a slight increase in the bioassay completion times. Therefore, the proposed general-purpose biochip offers high flexibility when running various bioassays on a dedicated layout.

Although Chips $A - D$ and Chips $E - G$ are bioassay-specific biochip, we still can "force" them to execute various bioassays by applying the scheduling algorithm proposed in Sect. 6.6. The execution times for running multiplexed bioassay, PCR bioassay, and protein dilution bioassay on Chips $A - D$ can be found in Table 6.6. The execution times for running multiplexed bioassay, PCR bioassay, and protein dilution bioassay on Chips $E - G$ can be found in Table 6.7.

By comparing the execution time in Tables 6.5, 6.6 and 6.7, we can find that Chips 1-4 can achieve shorter execution time comparing with Chips $A - D$ and Chips $E - G$ in most scenarios. This is because due to deadlocks, some operations that are implemented in parallel on Chips 1-4 have to be scheduled for sequential implementation on Chips $A - D$ and Chips $E - G$. Therefore, compared with the bioassay-specific biochip in [9], the proposed design has higher flexibility when running various bioassays on the same layout. Compared with the proposed general-purpose biochip, the ILP-based design method can reduce the pin-count by 33.3 % to 69.4 %. We have noted that there is a trade-off between pin-count and flexibility

Table 6.7 Execution time of bioassays on Chips $E - G$

Bioassay \ Biochip	Chip E	Chip F	Chip G
Multiplexed bioassay	87 s	131 s	NA
PCR bioassay	NA	20 s	NA
Protein bioassay	375 s	566 s	161 s

a

20	3	15	23	2	1	3	4
31	9	28	24	5	6	7	8
32	26	19	14	4	9	10	11
22	23	6	21	12	13	2	14
4	16	29	25	15	8	16	17
27	20	11	26	1	18	19	3
28	30	12	24	10	20	5	13
25	7	18	27	17	21	22	11

b

43	5	24	35	9	34	23	21	16	9
16	32	42	40	38	18	30	6	11	27
30	36	13	20	29	31	12	17	24	25
4	41	39	33	8	28	26	19	20	18
40	28	34	16	14	25	21	2	3	4
37	3	22	26	7	13	5	1	6	7
39	35	32	29	27	19	4	8	9	10
27	31	21	10	15	22	11	12	13	2
2	36	33	24	1	23	14	3	15	16
23	37	7	30	28	20	10	5	17	18

Fig. 6.19 (**a**) pin-assignment configuration for an 8×8 electrode array; (**b**) pin-assignment configuration for a 10×10 electrode array

when the bioassay-specific designs (Chips $A - D$ [9] and Chips $E - G$ [18]) and general purpose designs (Chips 1–4) are compared.

6.7.3   Simulation Results on Regular Array

Next we simulate bioassays on an 8×8 and a 10×10 electrode array to evaluate the performance of the proposed general-purpose biochip. The pin-assignment configurations of the 8×8 and 10×10 biochip are shown in Fig. 6.19a and 6.19b, respectively. There are 32 pins assigned to the 8×8 layout, and 43 pins assigned to the 10×10 layout. The numbers of control pins for the 8×8 direct-addressing and cross-referencing biochips are 64 and 16; the numbers of control pins for the 10×10 direct-addressing and cross-referencing biochips are 100 and 20, respectively.

The completion times of executing bioassays on the direct-addressing biochips, the cross-referencing biochips, and the proposed general-purpose biochips can be found in Table 6.8. Compared to the direct-addressing biochips, the general-purpose biochips can reduce the number of control pins by 50% or more with only a slight increase in the bioassay completion times.

Table 6.8 shows that for the protein dilution bioassay, the execution time on the 8×8 cross-referencing biochip is 20 % longer than that for the proposed 8×8 pin-limited biochip; for multiplexed bioassay and PCR bioassay, the execution times on these two biochips are almost the same. For the 10×10 array shown in Table 6.8, the cross-referencing biochip leads to lower completion time for the protein assay.

Table 6.8 Execution time of bioassays mapped to an 8×8 (10×10) electrode array

Biochip / Bioassay	Multiplexed	PCR	Protein
Direct-addressing biochip	72 s (72 s)	20 s (20 s)	160 s (142 s)
Cross-referencing biochip [6]	72 s (72 s)	20 s (20 s)	215 s (142 s)
Proposed general-purpose biochip	72 s (72 s)	24 s (20 s)	179 s (157 s)

Table 6.9 Comparison between lower bound, ILP model, and heuristic solution

Array size	No. pins (lower bound)	No. pins (ILP model)	No. pins (heuristic method)
3×3	9	9	9
5×7	18	21	23
7×8	24	29	30
8×8	26	32	32
9×9	30	Computationally impractical	40

However, cross-referencing biochips require special electrode structures (two electrodes per unit cell), which introduce fabrication challenges and additional processing steps [4]. As a result, the fabrication cost for cross-referencing biochip is likely to be higher than biochips without two electrodes per unit cell. For low-cost disposable chips, it is arguable whether cross-referencing design is always the right solution due to higher fabrication cost.

Finally, we present a comparison between the lower bound on the number of pins predicted by Theorem 6.1, the exact optimization results obtained using ILP model, and the results obtained using the heuristic procedure. Note that the lower bound on the number of pins may not always be achievable in practice and correspond to an actual pin-assignment configuration. On the other hand, the ILP model and the heuristic procedure can produce the configurations of the layouts. Hence the actual minimum number of required pins that we obtain by the ILP model is either equal to or larger than the theoretical lower bound.

Without loss of generality, we use various rectangular layouts as test cases. The results are shown in Table 6.9. The ILP model is solved using Xpress-MP [24], which is a widely used commercial ILP solver. The tool runs on a server with 64/,GB of memory and two quad-core 2.53/,GHz Intel Xeon processors. The computing time for the 8×8 array is about 5 h while for a 9×9 array, and the solver runs out of memory. The heuristic method is run on a 2.27 GHz Intel i3 Dual core processor with 2 GB of memory. The CPU times are less than 5 min. for all cases.

6.8 Chapter Summary and Conclusions

In this chapter, we have introduced a new method for mapping control pins to electrodes in the design of "general-purpose" (bioassay-independent) digital microfluidic biochips. We have derived necessary and sufficient conditions for pin-assignment to guarantee conflict-free concurrent manipulation of multiple droplets. The manipulation of droplets with different volumes on the biochip is considered. We have also presented a lower bound on the number of control pins required for an electrode array. A graph-theoretic "acceptance test" has been developed for a given pin-assignment. An ILP-based optimization technique has been described to automatically derive conflict-free pin-assignments. The ILP model is not scalable to large designs, but it can be used for partitioned designs and to evaluate the quality of heuristic solutions. We have presented an efficient heuristic approach for mapping control pins to electrodes. The heuristic method has been evaluated using commercial biochips and laboratory prototypes. Compared with previous techniques, the proposed method can reduce the number of control pins and facilitate the "general-purpose" use of digital microfluidic biochips for a wide range of applications.

References

1. Z. Hua, J. Rouse, A. Eckhardt, V. Srinivasan, V. Pamula, W. Schell, J. Benton, T. Mitchell, and M. Pollack, "Mutiplexed real-time polymerase chain reaction on a digital microfluidic platform", *Anal. Chem.*, vol. 82, pp. 2310–2316, 2010.
2. R. Sista, Z. Hua, P. Thwar, A. Sudarsan, V. Srinivasan, A. Eckhardt, M. Pollack, and V. Pamula, "Development of a digital microfluidic platform for point of care testing", *Lab on a Chip*, vol. 8, pp. 2091–2104, 2008.
3. (Silicon Biosystems) http://www.siliconbiosystems.com/
4. K. Chakrabarty, R. Fair and J. Zeng, "Design tools for digital microfluidic biochips: Towards functional diversification and more than Moore" (Keynote Paper), *IEEE Trans. CAD*, vol. 29, pp. 1001–1017, 2010.
5. V. Srinivasan, V. Pamula, and R. Fair, "An integrated digital microfluidic lab-on-a-chip for clinical diagnostics on human physiological fluids", *Lab on a Chip*, vol. 4, pp. 310–315, 2004.
6. Z. Xiao and E. Young, "CrossRouter: A droplet router for cross-referencing digital microfluidic biochips", *IEEE/ACM Asia South Pacific Design Automation Conference*, pp. 269–274, 2010.
7. T. Xu and K. Chakrabarty, "Broadcast electrode-addressing for pin-constrained multi-functional digital microfluidic biochips", *Proc. IEEE/ACM Design Automation Conference*, pp. 173–178, 2008.
8. C.-Y. Lin and Y.-W. Chang, "Cross-contamination aware design methodology for pin-constrained digital microfluidic biochips", *IEEE Transactions on Computer-Aided Design of Integrated Circuits and Systems*, vol. 30, pp. 817–828, No. 6, 2011.
9. Y. Zhao, T. Xu, and K. Chakrabarty, "Broadcast electrode-addressing and scheduling methods for pin-constrained digital microfluidic biochips", *IEEE Transactions on Computer-Aided Design of Integrated Circuits and Systems*, vol. 30, Issue 7, pp. 986–999, 2011.
10. F. Su and K. Chakrabarty, "Defect tolerance for gracefully-degradable microfluidics-based biochips", *Proc. IEEE VLSI Test Symposium*, pp. 321–326, 2005.

11. J. Park, S. Lee, and L. Kanga, "Fast and reliable droplet transport on single-plate electrowetting on dielectrics using nonfloating switching method", *Biomicrofluidics*, vol. 4, Issue. 2, pp. 1–8, 2010.
12. F. Harary, *Graph theory*, Addison-Wesley, MA: Reading, 1994.
13. Y. Zhao and K. Chakrabarty, "Simultaneous optimization of droplet routing and control-pin mapping to electrodes in digital microfluidic biochips", *IEEE Transactions on Computer-Aided Design of Integrated Circuits and Systems*, vol. 31, pp. 242–254, February 2012.
14. Y.-L. Hsieh, T.-Y. Ho, and K. Chakrabarty, "Design methodology for sample preparation on digital microfluidic biochips", *Proceedings of IEEE International Conference on Computer Design*, pp. 189–194, 2012.
15. H. Song, R. Evans, Y.-Y. Lin, B.-N. Hsu, and R. Fair, "A scaling model for electrowetting-on-dielectric microfluidic actuators", *Microfluidics and Nanofluidics*, vol. 7, Issue 1, pp. 75–89, 2009.
16. W. Nelson and C. J. Kim, "Droplet actuation by Electrowetting-on-Dielectric (EWOD): A review", *Journal of Adhesion Science and Technology*, Issue 26, pp. 1747–1771, 2012.
17. L. Jang, C. Hsu, and C. Chen, "Effect of electrode geometry on performance of EWOD device driven by battery-based system", *Biomed Microdevices*, Issue 11, pp. 1029–1036, 2009.
18. C.-Y. Lin and Y.-W. Chang, "ILP-based pin-count aware design methodology for microfluidic biochips", *IEEE Transactions on Computer-Aided Design of Integrated Circuits and Systems*, volume 29, Issue 9, pp. 1315–1327, 2010.
19. Y. Zhao, *Unified design and optimization tools for digital microfluidic biochips*, PhD Thesis, Duke University, Durham, NC, 2011.
20. Advanced Liquid Logic, http://www.liquid-logic.com.
21. D. Grissom and P. Brisk, "Path scheduling on digital microfluidic biochips", *IEEE/ACM Design Automation Conference*, pp. 26–35, 2012.
22. K. Chakrabarty and F. Su, *Digital Microfluidic Biochips: Synthesis, Testing, and Reconfiguration Techniques*, CRC Press, FL: Boca Raton, 2006.
23. L. Luo and S. Akella, "Optimal scheduling of biochemical analyses on digital microfluidic systems", *IEEE Transactions on Computer-Aided Design of Integrated Circuits and Systems*, vol. 8, Issue 1, pp. 216–227, 2011.
24. http://www.fico.com/en/Products/DMTools/xpress-overview/Pages/Xpr\discretionary-ess-Mosel.aspx

Chapter 7
Pin-Limited Cyberphysical Microfluidic Biochip

7.1 Introduction

On a digital microfluidic biochip, the number of control pins used to drive electrodes is a major contributor to the fabrication cost [1]. The layout of a typical biochip is shown in Fig. 7.1a, it has three regions:

1. **Active region**, where the biochemistry assays are executed; electrodes and on-chip reservoirs are fabricated in this region.
2. **Contact region**, where the contact pads of the input pins are fabricated. Here each contact pad corresponds to one input pin of the biochip. In order to reduce the contact resistance, the area of each pad is usually larger than an electrode [1].
3. **Wire routing region**, where metal wires are fabricated. The wires connect electrodes to the contact pads of the input pins.

From Fig. 7.1a, it is evident that most area of the biochip is occupied by the connection wires and contact pads. Therefore, the area and the fabrication cost of a biochip will increase when the number of the input pins increases.

In order to reduce the number of input pins in biochips, several design methods for pin-limited biochips have been presented in the literature. These design methods map a larger number of electrodes to a small number of control pins according to: (1) the specific synthesis result of a bioassay (i.e., bioassay-specific design for pin-assignment configurations) [2], or (2) the arrangement of electrodes on the biochip (i.e., application-independent design for pin-assignment configurations) [3–5].

In particulary, the design method proposed in [3, 4] generates an application-independent pin-assignment configuration with a minimum number of control pins. Compared with the bioassay-specific pin-assignment algorithms [2], the method in [3, 4] can reduce the number of control pins and facilitate the "general-purpose" use of digital microfluidic biochips for a wider range of applications; each general-purpose pin-limited biochip can be used to perform various target bioassays.

© Springer International Publishing Switzerland 2015
Y. Luo et al., *Hardware/Software Co-Design and Optimization for Cyberphysical Integration in Digital Microfluidic Biochips*, DOI 10.1007/978-3-319-09006-1_7

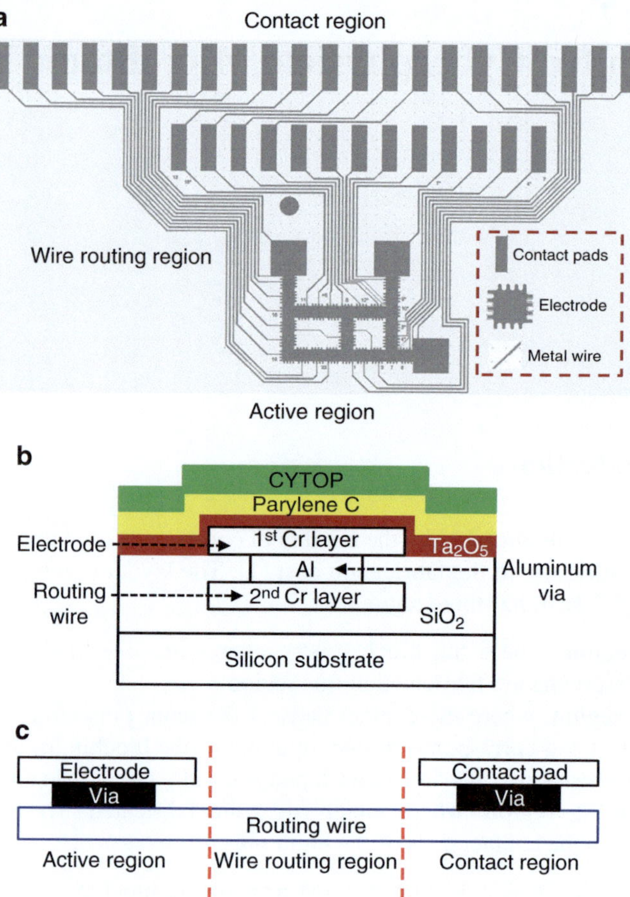

Fig. 7.1 (**a**) Layout of a biochip. The biochip has three regions: the region for fabricating electrodes and on-chip reservoirs, the region for wire routing, and the region for fabricating contact pads of input pins (Figure courtesy: Bang-Ning Hsu, Duke University); (**b**) side view of the substrate and bottom layer for the two-metal-layer biochip [6, 7]; (**c**) side view of the interconnection between an electrode and its corresponding contact pad

In a cyberphysical microfluidic system, the control software may have to dynamically adjust the schedules of fluidic operations as well as the routing paths of droplets. Different bioassays can execute on the same biochip, general-purpose pin-limited biochips [3, 4] are good candidates as hardware platforms for low-cost cyberphysical microfluidic systems.

However, despite the flexibility of general-purpose pin-limited biochips, the routability of metal wires in such chip has remained an open problem until now [3, 4].

In Sect. 7.2, we first discuss the wire-routing solution for general-purpose pin-limited biochips based on the two-metal-layer fabrication technique developed in recent years [6, 7]. Next, we discuss the specific design flow for pin-limited cyberphysical biochips in Sect. 7.3. Simulation results are presented in Sect. 7.4.

7.2 Wire Routing for General-Purpose Pin-Limited Biochips

For a biochip fabricated using the two-metal-layer technique of [6, 7], the side view of the substrate and the lower plate are shown in Fig. 7.1b. It is important to note that, increasing the number of metal layers will result in dramatic increases in the complexity and fabrication costs of the biochip, and its reliability will degrade sharply [8, 9]. Thus far, all fabricated silicon-based digital microfluidic biochips have no more than two metal-layers.

In the active region of the layout, the biochip has one metal layer for fabricating electrodes (i.e., the first Chromium layer shown in Fig. 7.1b), and another metal layer for wire routing (i.e., the second Chromium layer shown in Fig. 7.1b); these two metal layers are isolated by SiO_2. Under each electrode, there is an aluminum via that connects the electrode to its corresponding routing wire.

Similarly, in the contact region of the layout, the first metal layer is used to fabricate the contact pads of the input pins. For each contact pad, there is an aluminum via that connects the contact pad to the corresponding routing wire (which is fabricated by the second metal layer). The side view of the interconnection between the electrode and the corresponding contact pad is shown in Fig. 7.1c. By using the two-metal-layer fabrication technique, electrodes in a large array can be connected to their corresponding input pins.

Assume that we have already derived the pin-assignment configuration for a $W \times H$ electrode array, and the electrodes are mapped to P pins. Based on the existing two-metal-layer fabrication technique [6, 7], the wire-routing problem for biochip can be considered as an example of a constraint-satisfaction problem, i.e., how to derive a feasible wire-routing solution for any given pin-assignment configuration using no more than two metal layers on the silicon-based biochip. In order to solve this problem, a constructive method for wire-routing is proposed. The wire routing solution of a pin-limited biochip can be derived using the following steps.

1. **Adding "routing rings" in the wire routing region.** In the wire routing region, "routing rings" are fabricated using the first metal layer, as shown in Fig. 7.2a. Each routing ring is assigned to one control pin of the biochip, and all the routing rings are axis-parallel rectangles. The total number of fabricated routing rings is P. The edges of any two rectangles do not intersect, and the routing rings are electrically isolated from each other.
2. **Partitioning an electrode into four parts.** A large electrode array can be "equally" partitioned into upper left, upper right, lower left, and lower right sub-arrays, as shown in Fig. 7.2b. The wire routing for these four sub-arrays are derived separately. In the active region, routing wires in different sub-arrays do not interference with each other.

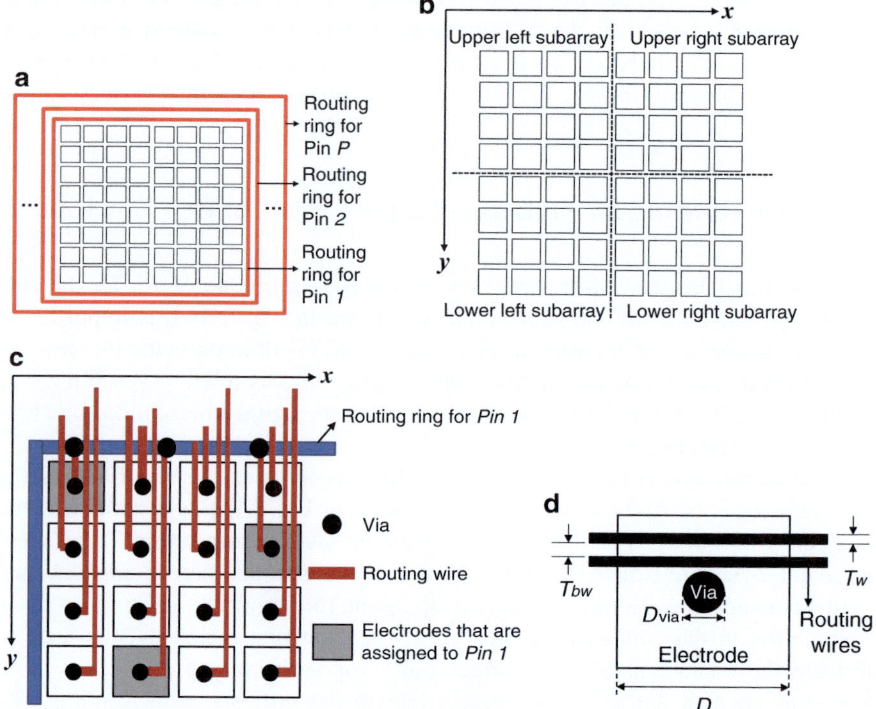

Fig. 7.2 (**a**) An electrode array and "routing rings" for Pins 1 to P; (**b**) a large electrode array that is partitioned into four parts. The wire routing solution for these parts can be derived separately; (**c**) *top view* of an electrode sub-array and the underneath routing wires on the direction which is parallel with one edge of the array; (**d**) *top view* of an electrode and the underneath routing wires

3. **Connecting electrodes to routing rings.** Based on the pin-assignment configuration, electrodes are connected to their corresponding routing rings. An example is as follows. We assume that the electrode array shown in Fig. 7.2c is the upper left sub-array of the electrode array shown in Fig. 7.2b. In this sub-array, three electrodes are assigned to be connected to Pin 1. The vias at the centers of these three electrodes electrically connect the electrodes to three routing wires. As shown in Fig. 7.2c, routing wires of these three electrodes intersect with the routing ring for Pin 1. At these crosspoints, vias are added to electrically connect the routing wires with the routing ring for Pin 1. Using this method, all the electrodes that are assigned to Pin 1 are connected to the routing ring for Pin 1.

In the similar manner, all electrodes that are assigned to Pin p ($1 \leq p \leq P$) are connected to the routing ring for Pin p. In order to simplify the wire-routing solution and to avoid any interference between two metal wires on the two-metal-layer biochip, all routing wires are aligned parallel to the edge of the electrode array (in either the direction of x-axis or y-axis), as shown in Fig. 7.2c.

4. **Connecting routing rings to input pads.** In the contact region, P contact pads are fabricated using the first metal layer, and each routing ring is connected to its corresponding contact pad. The connection wires are fabricated using the second metal layer.

Among the above steps, we note that the routability of a pin-limited biochip is determined in Step 3, i.e., if there is enough space underneath each electrode to place the parallel routing wires, the pin-limited biochip is routable. As shown in Fig. 7.2c, at the edge of the sub-array that is perpendicular to the routing wires, the electrodes have the maximum number of routing wires underneath them (i.e., the four electrodes at the first row of the 4×4 sub-array each has four routing wires underneath it). Similarly, for a $W_s \times H_s$ ($W_s \neq H_s$) rectangular sub-array, if the routing wires are aligned parallel to the longer edge of the rectangle, the maximum number of routing wires fabricated underneath the electrodes is $\max\{W_s, H_s\}$; if the routing wires are aligned parallel to the shorter edge of the rectangle, the maximum number of routing wires fabricated underneath the electrodes is $\min\{W_s, H_s\}$. As the orientation of routing wires can be parallel to either the longer or shorter edge of the sub-array, we will always align routing wires parallel to the shorter edge of the rectangle when designing the wire-routing solution for the $W_s \times H_s$ rectangular sub-array. Hence, the maximum number of routing wires fabricated underneath electrodes in the sub-array is $\min\{W_s, H_s\}$.

Therefore, based on the size of the electrode array, as well as the geometric sizes of electrodes, routing wires, and vias, we can determine whether a pin-limited biochip has a feasible wire routing solution. For a $W \times H$ pin-limited biochip, the sufficient condition for the existence of its wire routing solution can be derived as follows.

The top view of an electrode and the underneath routing wires are shown in Fig. 7.2d. Here we assume that the diameter of the via is D_{via}, the width and length of each electrode are both D, and the width of a routing wire is T_{w}. In order to prevent electrical shorting of the two wires, the gap between them is T_{bw}. Therefore, the average "occupied" width of each routing wire can be estimated as $(T_{\text{w}} + T_{\text{bw}})$.

As shown in Fig. 7.1b and Fig. 7.2d, there is a via at the center of each electrode. Since the routing wire cannot overlap with the via, the "effective width" under each electrode that can be used for aligning routing wires is estimated as $(D - D_{\text{via}})$. Therefore, the maximum number of routing wires that can be fabricated underneath an electrode can be estimated as $\lfloor \frac{D - D_{\text{via}}}{T_{\text{w}} + T_{\text{bw}}} \rfloor$, where $\lfloor * \rfloor$ represents the largest integer not greater than "$*$".

As discussed above, a $W \times H$ electrode array can be "equally" partitioned into four sub-arrays, and the size of the maximum sub-array is $\lceil \frac{W}{2} \rceil \times \lceil \frac{H}{2} \rceil$, where $\lceil * \rceil$ represents the smallest integer not less than "$*$". Therefore, a $W \times H$ two-metal-layer biochip is routable when: $\min(\lceil \frac{W}{2} \rceil, \lceil \frac{H}{2} \rceil) \leq \lfloor \frac{D - D_{\text{via}}}{T_{\text{w}} + T_{\text{bw}}} \rfloor$.

The typical values for D, D_{via}, T_{w}, T_{bw} are $1{,}000\,\mu\text{m}$, $200\,\mu\text{m}$, $20\,\mu\text{m}$, and $20\,\mu\text{m}$, respectively [6,7]. Therefore, if $W \leq 40$ or $H \leq 40$, the pin-limited biochip fabricated by two-metal-layer technique is routable. For most of the benchmark bioassays that are extensively discussed in literature, the sizes of the required electrode arrays are smaller than 40×40. Hence for the commonly used pin-limited biochips, we can derive feasible wire routing solutions by Steps 1–4 listed above.

7.3 Design Flow for Pin-Limited Cyberphysical Biochips

For a given $W \times H$ electrode array without pin-assignment, the heuristic algorithm proposed in [3, 4] can first derive a pin-assignment configuration that is application-independent. Next, if $W \leq 40$ or $H \leq 40$, by applying Steps 1–4 proposed in Sect. 7.2, we always can find a feasible wire routing solution for this pin-limited biochip. For any given bioassay, by applying the operation-dependency-aware synthesis algorithm and the droplet-routing approach described in Sects. 4.3 and 4.4, the synthesis result of the bioassay on a $W \times H$ direct-addressing biochips can be derived. The synthesis result is defined as the "initial synthesis results".

Next, the scheduling algorithm proposed in [4] is used to adjust the "initial synthesis result" (derived on a direct-addressing biochip) according to the pin-assignment configuration of the layout. Finally, the synthesis results of the bioassay on the pin-limited biochip can be derived, i.e., the bioassay can now be executed on the low-cost pin-limited biochip. The flowchart for the complete design flow is shown in Fig. 7.3. All the information required for the execution of the bioassay (including the scheduling and module placement of operations; transportation paths of droplets; the pin-assignment configuration and wire-routing solution for the biochip) can be derived in the design flow.

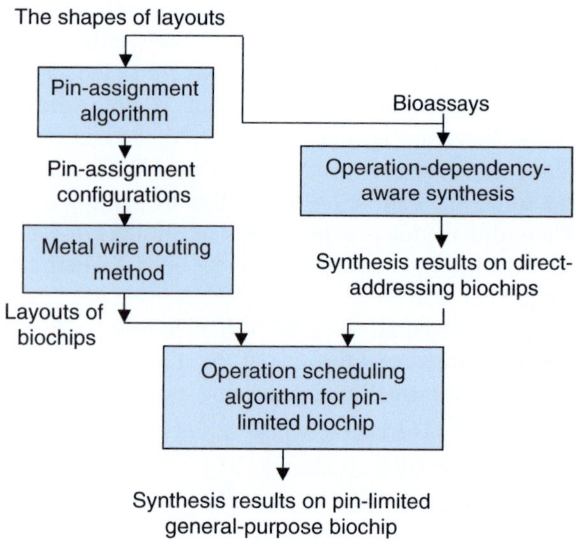

Fig. 7.3 The design flow for pin-limited cyberphysical biochips

7.4 Simulation Results

7.4.1 Results Derived by the Operation-Interdependency-Aware Synthesis Approach

In Sect. 4.5.3, we have mapped bioassays to direct-addressing biochips with different sizes. The corresponding completion times of bioassays in dilution/mixing (D/M) phases and transportation (T) phases on the direct-addressing biochips are shown in Table 4.5.

As discussed in Sect. 6.6, we can map bioassays to pin-limited general purpose biochips proposed in [3, 4]. The pin-assignment configurations for 8×8, 9×9, 10×8, and 12×8 electrode arrays are shown in Fig. 7.4. Compared with the direct-addressing scheme, the design of pin-limited general-purpose biochips can reduce the pin-counts by 50.0–65.8 %.

The synthesis results for executing bioassays on direct-addressing biochips are defined as "initial results". By applying the scheduling algorithm proposed in [4] on the initial results, we derive the synthesis results for bioassays. Hence the bioassays can be mapped to the general-purpose pin-limited biochips shown in Fig. 7.4. The corresponding completion times of bioassays in D/M phases and T phases on these pin-limited general-purpose biochips are shown in Table 7.1. The clock frequency of the T phase is also set as 1 Hz.

a

20	3	15	23	2	1	3	4
31	9	28	24	5	6	7	8
32	26	19	14	4	9	10	11
22	23	6	21	12	13	2	14
4	16	29	25	15	8	16	17
27	20	11	26	1	18	19	3
28	30	12	24	10	20	5	13
25	7	18	27	17	21	22	11

b

18	36	20	26	29	21	2	3	4
3	25	28	9	16	5	1	6	7
37	32	21	19	22	4	8	9	10
27	6	30	17	25	11	12	13	2
36	34	31	2	26	14	3	15	16
12	16	20	23	27	10	5	17	18
24	35	32	11	15	19	20	7	8
30	7	33	29	24	1	13	14	21
37	22	12	6	28	18	22	23	24

c

32	2	37	34	8	30	25	19	21	22
27	38	17	33	29	22	23	2	3	4
10	30	1	31	27	13	5	1	6	7
21	35	11	15	19	24	4	8	9	10
31	9	22	32	6	18	11	12	13	2
5	16	34	20	28	25	14	3	15	16
37	36	14	1	29	26	10	5	17	18
28	7	24	21	16	12	19	20	7	8

d

31	7	41	37	39	3	33	17	23	26	1	20
24	42	40	17	11	35	19	30	4	22	21	16
26	28	5	20	34	31	10	25	2	15	17	9
6	39	29	36	30	1	27	24	20	18	12	6
40	38	31	32	13	28	12	22	19	7	8	2
35	12	4	16	18	25	11	5	13	9	3	1
14	37	33	27	29	21	23	8	14	10	4	5
2	32	34	26	14	3	24	16	15	11	6	7

Fig. 7.4 Pin-assignment configuration for: (**a**) 8×8 array; (**b**) 9×9 array; (**c**) 10×8 array; and (**d**) 12×8 array

Table 7.1 Synthesis results derived by the operation-interdependency-aware synthesis approach on pin-limited biochips

Bioassay	Size of the biochip	Completion time (s)		
		D/M phases	T phases	Total
Exponential dilution bioassay	8×8	116	181	297
	9×9	59	163	222
	10×8	67	161	228
	12×8	54	244	298
Interpolation dilution bioassay	8×8	89	162	251
	9×9	45	133	178
	10×8	48	165	213
	12×8	45	189	234

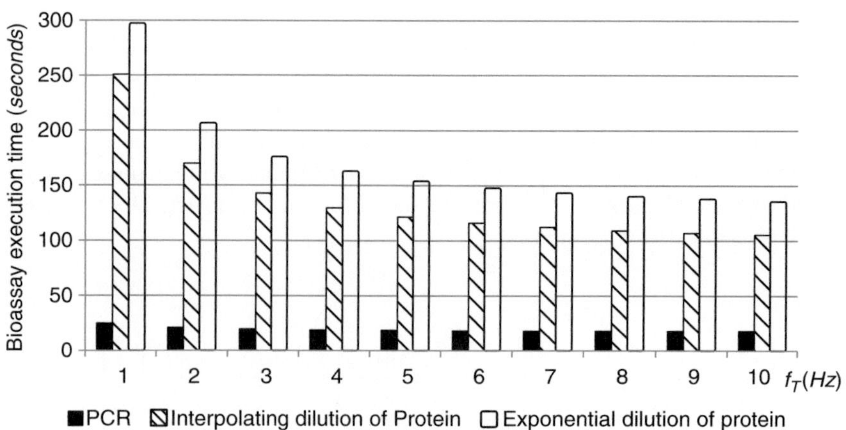

Fig. 7.5 The relationship between completion time of bioassays and f_T on (**a**) an 8×8 direct-addressing biochip; (**b**) the 8×8 biochip shown in Fig. 7.4a

7.4.2 Completion Time with Multiple Clock Frequencies on Pin-Limited Biochip

In Sect. 4.5.4 we present the relationship between the bioassay completion time and the clock frequency of the droplet transaction phase on a biochip with multiple working frequencies. If the bioassays are executed on the 8×8 pin-limited electrode array shown in Fig. 7.4a, the relationship between completion time of bioassays and the clock frequency of the T phase is shown in Fig. 7.5. The execution times for exponential dilution and interpolation dilution bioassays on the pin-limited biochip can be greatly reduced by increasing f_T. For example, when f_T increases from 1 Hz to 10 Hz, the execution time for exponential dilution bioassay is reduced from 297 s to 134 s.

From the simulation results, we find that when the clock frequency of the T phase increases, the completion time of PCR bioassay on the pin-limited biochip does not decrease significantly. This is because the time for droplet transportation in the PCR bioassay is relatively low.

7.5 Chapter Summary and Conclusions

In this chapter, we discussed a solution for metal wire routing on pin-limited biochip. The proposed method can guarantee that with a two-metal-layer fabrication technique, all the widely-used pin-limited biochips are routable. In order to reduce the cost of the cyberphysical biochip, we also proposed the design flow for pin-limited cyberphysical microfluidic biochip. The simulation results show that, the proposed method can reduce the number of control pins and facilitate the "general-purpose" use of cyberphysical digital microfluidic biochips for a wide range of applications.

References

1. K. Chakrabarty, R. Fair, and J. Zeng, "Design tools for digital microfluidic biochips: Towards functional diversification and more than Moore" (Keynote Paper), *IEEE Trans. CAD*, vol. 29, pp. 1001–1017, 2010.
2. C.-Y. Lin and Y.-W. Chang, "ILP-based pin-count aware design methodology for microfluidic biochips", *IEEE Transactions on Computer-Aided Design of Integrated Circuits and Systems*, volume 29, Issue 9, pp. 1315–1327, 2010.
3. Y. Luo and K. Chakrabarty, "Design of pin-constrained general-purpose digital microfluidic biochips", *ACM/IEEE Design Automation Conference*, pp. 18–25, 2012.
4. Y. Luo and K. Chakrabarty, "Design of pin-constrained general-purpose digital microfluidic biochips", *IEEE Transactions on Computer-Aided Design of Integrated Circuits and Systems*, Volume 32, Issue 9, pp. 1307–1320, 2013 (http://ieeexplore.ieee.org/xpl/articleDetails.jsp?arnumber=6582577).
5. Z. Xiao and E. Young, "CrossRouter: A droplet router for cross-referencing digital microfluidic biochips", *IEEE/ACM Asia South Pacific Design Automation Conference*, pp. 269–274, 2010.
6. Y.-Y. Lin, E. Welch, and R. Fair, "Low voltage picoliter droplet manipulation utilizfing electrowetting-on-dielectric platforms", *Sensors and Actuators, B: Chemical*, vol. 173, pp. 338–345, 2012.
7. J. Gao, X. Liu, T. Chen, P.-I. Mak, Y. Du, M. Vai, B. Lin, and R. Martins, "An intelligent digital microfluidic system with fuzzy enhanced feedback for multi-droplet manipulation", *Lab on a Chip*, Issue 13, volume 13, 2013.
8. T. Xu and K. Chakrabarty, "Droplet-trace-based array partitioning and a pin assignment algorithm for the automated design of digital microfluidic biochips", *Proc. IEEE/ACM International Conference on Hardware/Software Codesign and System Synthesis*, pp. 112–117, 2006.
9. T.-W. Huang, J.-W. Chang, and T.-Y. Ho, "Integrated fluidic-chip co-design methodology for digital microfluidic biochips", *Proceedings of ACM International Symposium on Physical Design*, pp. 49–56, 2012.

Chapter 8
Conclusions

This book has described a set of design techniques for cyberphysical microfluidic biochips. Contributions include design methods that consider timing uncertainties in operations, optimization to handle droplets with various volumes, layout optimization of cyberphysical PCR biochips, and a design flow for cyberphysical pin-limited microfluidic biochips. Compared to prior design methods, this research work has led to a computer-aided design flow that considers physical constraints and enables the development of cyberphysical biochip systems.

We have presented the first step towards physical-aware optimization and control software design. We have developed an algorithm for droplet tracking based on pictures captured by CCD cameras, and a reliability-driven error recovery strategy. Using feedback from sensors, the control software can determine whether there are errors on the biochip. The control software discards the region where an error has occurred to ensure reliable chip operation. We have also developed a re-synthesis method for cyberphysical biochip, which can dynamically alter the schedule of fluid-handling operations based on sensors' feedback.

An important contribution of this book includes a dictionary-based error recovery method for low-cost portable cyberphysical microfluidic systems. In order to derive error dictionaries, a technique has been developed for realistic fault simulation. By utilizing error dictionaries, we can achieve a cost-effective implementation of cyberphysical systems to recover from errors that occur during chip usage.

In order to mimic the behavior of faulty fluid-handling operations, we have developed a fault simulation framework that can integrate data collected from real experiments. In prior work, error recovery and synthesis were based on the assumption that all operations have the same failure probability, but this assumption may not reflect reality. In order to address this problem, we have presented failure models of biochips based on real data, and used these models to guide fault simulation.

© Springer International Publishing Switzerland 2015

Y. Luo et al., *Hardware/Software Co-Design and Optimization for Cyberphysical Integration in Digital Microfluidic Biochips*, DOI 10.1007/978-3-319-09006-1_8

By considering the physical constraints in cyberphysical microfluidic system, we have developed a layout optimization technique for cyberphysical PCR biochips. A statistical model for DNA amplification has been introduced. The module is used to predict whether the droplet has enough DNA strands to carry out the PCR, and it helps us to improve the efficiency of PCR on cyberphysical biochips. We have developed a layout algorithm that considers interference between the on-chip elements such as reservoirs, sensors and thermal units. An optimization algorithm has been developed to minimize the area and electrode count of the biochip and satisfy proximity constraints. Several real-life bioassays have been used to demonstrate that the proposed design method reduces the overall mixing/dilution/detection time for a given bioassay.

This book has also presented a synthesis algorithm for bioassays that is capable of taking into account uncertainties inherent in biochemical reactions. The proposed time-uncertainty aware synthesis algorithm can overcome the inherent variability and randomness of biological/chemical processes and improve the quality of product droplets. An "operation-interdependence-aware" synthesis algorithm has been developed—the first on-chip biochemistry synthesis procedure that does not use the module library as a design guideline. Using this algorithm, the dependence on characterization can be reduced. The algorithm leads to a design approach that considers completion-time uncertainties for fluidic operations, hence the accuracy of fluidic operations is also improved. The proposed approach explicitly considers the transportation of droplets. Hence it guarantees the feasibility of droplet-routing, and determines the transportation path for each droplet. In order to reduce the execution time of the bioassay without additional degradation of electrodes or hardware cost, the design of microfluidic biochips using multiple clock frequencies has also been considered. An on-line droplet-routing method with low computational complexity has been developed. The response time of the resulting cyberphysical system is negligible. Therefore, the degeneration of intermediate products in the bioassay can be avoided.

To minimize the number of control pins and simultaneously maximize the flexibility of biochips, we have developed design techniques for general-purpose pin-limited biochips. A graph model for analyzing pin-assignment configurations of a biochip has been developed. Based on the graph model, we have developed a heuristic algorithm for the design of pin-assignment configuration, devised an acceptance test, and developed a scheduling algorithm for fluid-handling operations. The manipulation of droplets with various volumes has also been considered, and new design constraints associated with the processing of large droplets have been discussed. The metal wire routing method for pin-limited has been investigated. Finally, by combining the design algorithms for general-purpose pin-limited biochip and cyberphysical microfluidic biochips, a design flow for cyberphysical pin-limited biochips has been developed.

In summary, this book has presented a comprehensive algorithmic infrastructure for the design and optimization of a cyberphysical biochip system. A series of related design problems for cyberphysical microfluidic systems have been addressed, including uncertainty-aware synthesis, dynamic error recovery, layout optimization, pin-count minimization, and optimization for cyberphysical pin-limited biochips. This work has therefore led to powerful design tools for cyber-physical microfluidic biochips and can serve as a bridge between theoretical design algorithms and realistic applications of biochips.